Power Electronics: DC-DC Converters for HVDC Transformations

By

Hemchandra M. Shertukde, Ph.D., P.E.

DDI Publishing Group

Diagnostic Devices Inc. dba DDI PO Box 362,

11 Windham Drive, Simsbury, CT, 06070, USA

ISBN (Paperback): 979-8-3493-9702-8

Edited and Published by Callaghan Publications –

CALLAGHAN
PUBLICATIONS

Visit the DDI Publishing Web site at: www.diagnostic-devices.com

Table of Contents

Preface

My sincere wish to write a book on DC-DC Converters for HVDC Transformations came to fruition when the University of Hartford awarded me a Special Sabbatical for AY 20-21.

Earlier, this same opportunity was afforded by the University of Hartford by awarding me a regular Sabbatical leave for AY 2019-20. However, the publisher, in this case Momentum Press folded its business and so there was no publisher left to publish this book on this topic. I approached other publishers, but none agreed till CRC Press agreed to publish this book and awarded me a contract to write this book on DC-DC Converters for HVDC Transformations. Thanks to Mr. Marc Gutierrez, Editor – Electrical & Biomedical Engineering | CRC Press | Taylor & Francis Group, for making this possible. Incidentally, the previous Chief Publisher of Momentum Press, Mr. Joel Stein then became acquisition Editor for CRC Press and helped in making this project possible as well. CRC reneged and the publication was finally taken up by Callaghan Publishing Company an arm of Amazon Publishers.

I have made an earnest effort to present material from the basic circuit configurations of different DC-DC Converters at normal LV voltages and extending it to HVDC configurations. Different configurations like buck, boost, buck-boost DC-DC converters, SEPIC converters, Cook', Fly back' etc., are handled. Basic devices for such applications as normal power diodes. MOSFETS and IGBT/IGCTs are described in different LVDCs. MVDC and HVDC applications.

This book is suitable for instructions at the Senior undergraduate level and Freshman Graduate level for a two-semester course in Power Electronics dedicated to DC-DC converters, both at the low voltage levels, i.e., LVDC, medium voltage levels, i.e., MVDC and high voltage levels, i.e. HVDC, as attempted in this book respectively. Through understanding of Buck-boost methodology with a case study to illustrate the design, and manufacture using Volterra PCB machines and surface mount components (smcs) is explained in this book.

Foreword

This book on DC-DC Converters for HVDC Transformations is a culmination of my efforts over the past two years in preparing the manuscript from a set of personal experiences in this area and after teaching this course over the past thirty years at the graduate level in the S.I. Ward Department of Electrical Engineering in the College of Engineering Technology and Architecture (CETA) at the University of Hartford. As the Director of Graduate Studies and after developing the Master of Engineering degree program for the college, I had initiated all the 'blue sheets' for most of the courses, Power Electronics ECE 573 being one of them.

Incidentally, I also taught this course last Fall. Additional material, like introducing new problems for each chapter, were formulated during this last Fall, and some of the graduate students liked the use of different software packages like PSpice and MuitiSim in solving such real-time problems for all types of DC-DC converters. Some of them are included in this version of the book. I would like to acknowledge the help of graduate student Esmaeil Gandomkar in this effort. This textbook should be used to illustrate practical cases in DC-DC Converters for HVDC Transformations as well.

I dedicate this book to my wife Rekha Shertukde and my elder sister Mrs. Anjali Kulkarni.

Acknowledgement

I want to acknowledge that this book is a result of the 'Special Sabbatical' awarded to me by the University of Hartford for the academic year 2020-2021. This was offered as a means for the University to save money due to the COVID-19 pandemic conditions, which threw the entire world into a panic mode. I consider this to be my small contribution to the prevention of a global disaster.

I am grateful for this opportunity, as it has given me one more chance to write a book on my favorite topic: Power Electronics – DC-DC Converters for HVDC Transformations, during this sagging economy and dropping enrollments at the University of Hartford.

I would also like to dedicate this book to the latest addition to our family:

Ryaan J Bhakta, who is my second grandchild, born October 4, 2020.

And;

Supriya Rohan Shertukde, our new daughter-in-law, who married my son Rohan on October 15, 2021.

Welcome to the Shertukde family.

Thank you to all my family members for their cooperation and support over the past three years since the actual manuscript writing started in the Fall of 2019.

Hemchandra M. Shertukde, Ph.D., P.E., DMS

Chapter 1

Introduction to DC-DC Converters for HVDC Transformation

The need for DC-DC converters for HVDC Transformation started in right earnest in the 20th century when electrical power transmission and distribution rose to above 765 KV. The need for the HVDC has taken a further leap in its application with the harnessing of alternate energy sources, mainly solar energy and off-shore wind farms. Solar energy is harnessed at 28-34 volts DC, and the next level of voltage in the transmission and distribution part of a power system grid rises to as high as 1.1 kV (or 1100 volts ac). Such a change from DC to AC is affected by Inverters. This change in voltage level then needs conversion of AC to dc, and thus a need for ac-dc converters, which are typically called rectifiers. Further, some operating and control devices need stepping up or stepping down of DC voltages to higher than 34 volts or lower than 34 volts; and thus, the need for dc-dc converters. Small dc-dc conversion from 5 volts to 15 volts is applicable in electronic circuitry, but to step up or boost voltage from 34 volts to 1100 volts in power circuits needs careful design engineering methodology with utmost attention provided to the filtering aspect of the eventual signal. In this book we will study the different components of the dc-dc converter in block diagram form with insights into the different nuances of buck and boost converters at high voltage levels. These could be as high as 765 kV and above. In a power system, one-line diagram (OLD), one encounters different levels of voltages as enumerated below:

1) Generation level: 22 kV to 33 kV

2) Transmission and Distribution level: from 1.1 kV to 1500 kV.

3) Distribution level at load level is as low as 115 volts to 440/230 volts up to 69 kV

The first two categories need boosting of voltages as well and the second and third categories need bucking of the voltage levels.

In addition, one can encounter another form of DC-DC converter called the Single-Ended Primary Input Converter (SEPIC). The essential component that differentiates the buck-boost converters from the SEPIC converters is the presence of a suitable inductor. All these converters are built in the High Voltage Direct Current (HVDC) domain using Integrated Gate Biased Transistor IGBTs and Integrated Gate-Coupled Transistor (IGCT)s. In the beginning of the design of such DC-DC Converters Silicon Controlled Rectifier (SCR) were generally used for lower power consumption. However, due to the large power requirement of HVDC systems, IGBTs and IGCTs find a better and more efficient application.

Power devices used in HVDC

1) Power Diodes

2) Power Transistors

3) Power Silicone Control Rectifier (SCR)

4) Metal Oxide Sulphur dioxide Field Effect Transistor (MOSFET)

5) Insulated Gate Biased Transistor (IGBT)

IGBTs are widely used in the low-voltage systems. IGBTs offer fast interruption time and the ability to withstand short-circuit currents, but have high conduction losses. These are currently in use for up to 6.3 kV, but laboratory development has been conducted at 8 kV since 2003, with considerable research dedicated by ABB, Switzerland.

6) Insulated Gate Coupled Transistor (IGCT)

IGCTs have the lower conduction losses of a thyristor with the turn-off capability of a transistor. These are currently in use for up to 6.3 kV, but laboratory development has been conducted at 8 kV since 2003 with considerable research dedicated by ABB, Switzerland. ABB has introduced a newer version called as the Soft Punch Through (SPT) – IGBT.

The SPT buffer region consists of a deep and low surface concentration doping profile, as shown in reference (2).

The use of the HVDC grid system is based on the economical consideration to costs involved in a two-wire electrical system to a four-wire electrical system. The saving in the number of conductors from four to two in the former saves the cost of copper for the entire grid, thus reducing the capital cost for the entire one-line diagram (OLD) as mentioned above.

The economics are significant and result in a saving of 25-30%.

The major cost of such a system is focused on the filters that are used to affect a regulated supply where the ripple factor in such systems is as low as < 0.1%. The tight ripple factor requirements increase the cost of the filtering system utilized in the designs of the high voltage and power level DC signals.

Cross Regulation Effects of DC-DC Converters

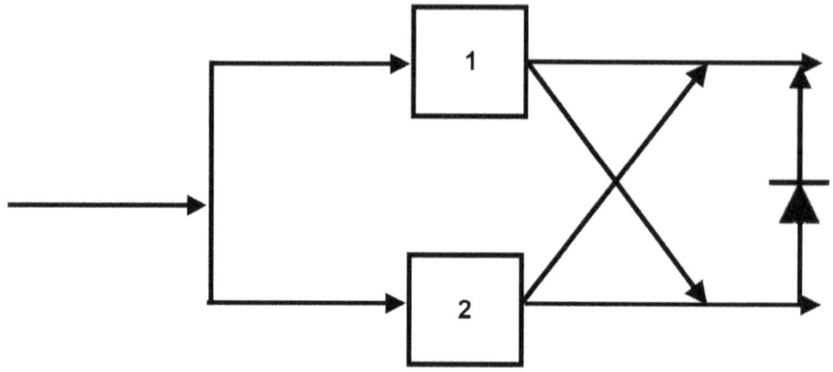

Figure 1 circuit for multiple converters connected in a cross-regulated condition.

This figure shows dual-output DC-DC Converters. The configuration consists of an isolation transformer with a single input winding and dual output windings, as shown. The positive output is from the top output winding, and the negative output is from the bottom output winding of the transformer. Note the dot markings on the output windings. The two outputs are connected as shown in the figure and connected to the voltage regulator and the control loop of the DC-DC Converter. The regulation is obtained by the Zener diode, as shown between the common point, and controls the cross-regulating effects of the load variation.

This is shown in the figure below. When the load increases, the negative voltage tends to move towards zero. The point of equilibrium is shown as the intersection of the bold line with the dashed line. Accordingly, when the load decreases, the voltage moves towards more negative voltage (i.e., like -13 volts).

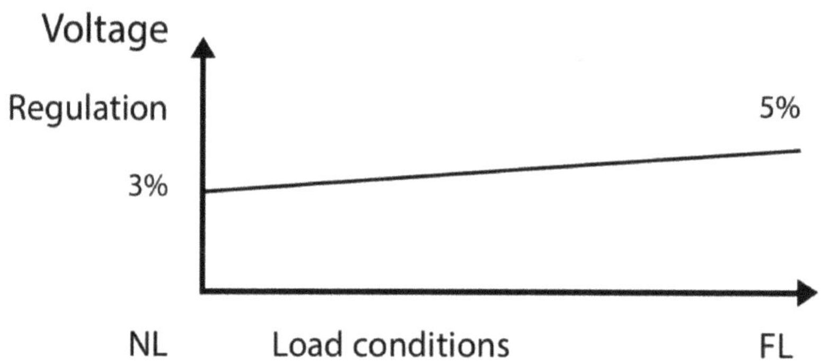

Figure 2 Voltage vs load conditions and respective % regulation.

The cross-regulation is specified at 50% in the figure above, in this case as 50%.

The horizontal axis is the percentage of negative load current. Now, say the dashed the line is surrounded by the 30% load, and the line below is 70% regulation, and the dashed line is the balance load condition intersecting the bold line. This balanced load is regulated by the Zener diode.

Chapter 1 Problems

Problem 1: The emitter follower, as in the figure we discussed in class, VCC = 15 V, $V_{CE}sat$ = 0.2 V and constant, and β is very high. R is in the biasing of base of Q2.

a) Find the value of R that will establish a bias current sufficiently large to allow the largest possible output signal swing for RL = 1 kΩ.
b) Determine the resulting output signal and the minimum and maximum emitter currents.

Problem 2: Consider the emitter follower as in Problem 1 above with V_{cc} = 10V, I = 100 mA and RL = 100 Ω.

a) Find the power dissipated in Q1 and Q2 under quiescent conditions (v0 = 0). For a sinusoidal output voltage of maximum possible amplitude (neglecting V_{CE} (sat)).

b) Find the average power dissipation in Q1 and Q2.

c) Also find the load power

Problem 3: For the emitter follower of Problem 1 let V_{cc} = 10V, I = 100 mA, and R_L = 100 Ω. If the output voltage is an 8 V peak.

a) Find the power delivered to the load.

b) The average power drawn from the supplies.

c) Power conversion efficiency. Ignore the loss in Q

Problem 4: Enhancement type NMOS transistor with V_i = 2V has its source terminal grounded and a 3 VDC source connected to the gate. In what region of operation does the device operate for:

a) V_D = + 0.5 V?

b) V_D = 1 V?

c) V_D = 3 V?

Problem 5: If the NMOS device in Problem 1 above has $\mu_{n}C_{oX}$ = 20μA/V2 , W =100 μm and L = 10 μm, find the value of drain current that results in each of the following three cases, as in Problem 4. Neglect the dependence of i_D on v_{DS} in saturation.

Chapter 2

Different Kinds of DC-DC Converter

Chapter 2 describes the different kinds of DC-DC Converters that are primarily encountered in DC-to-DC voltage conversion of DC voltage. Block diagrams of DC-DC converters are illustrated below in Figure 2.1 & 2.2.

Buck DC-DC Converters are used to reduce input DC voltage from a higher value to a lower magnitude DC voltage (i.e., 28 volts to 5 volts LVDC). While Boost DC-DC Converters increase voltage magnitude at the output or load end from the one at the source end (5 volts to 28 volts LVDC). The next category is called the Fly back DC-DC Converter, which is also called an isolation DC-DC Converter.

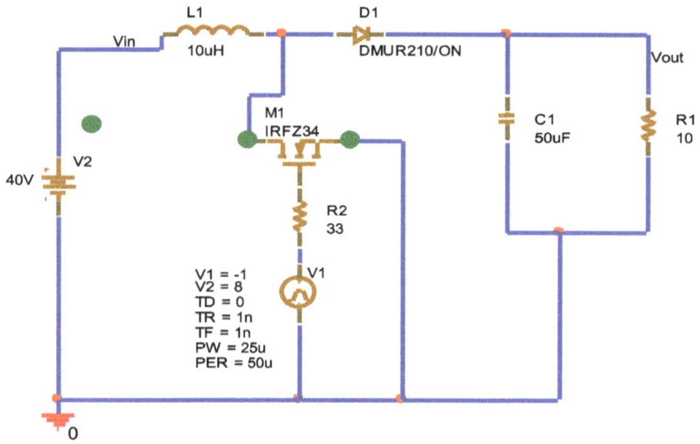

2.1 Typical Buck Converter

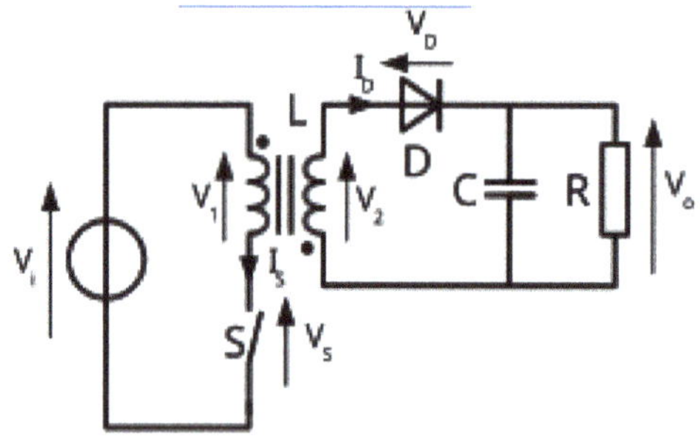

2.2 Typical Flyback

These are converters with isolation. Useful while operating dc-dc converters off the power to lines. This system of design is used in chargers and other devices connected directly to power lines. The name fly back comes from an analog TV application. Fly back circuitry is useful in bringing the beam of electrons back to the TV screen, thus the name Fly back.

At this time, we also need to understand what a coupled inductor is? This is different from a transformer. In a coupled inductor that has current coming in, energy is stored in the core and can get back this energy from another winding later at some point.

This is shown in equation 2.1 for a configuration as in Figure 2.3.

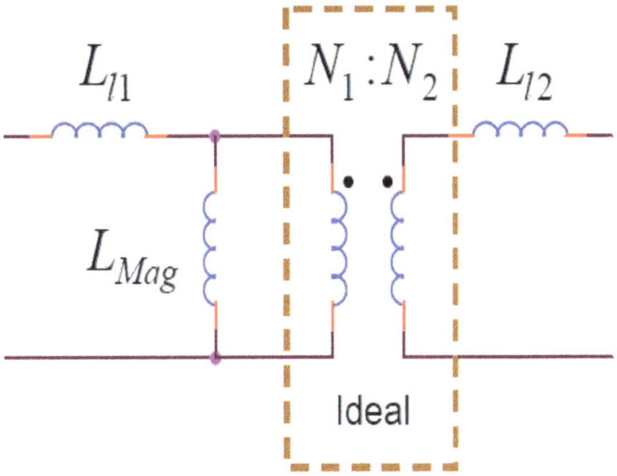

Figure 2.3 Coupled

$$\frac{L_1}{L_2} = \left(\frac{n_1}{n_2}\right)^2 \qquad 2.1$$

Figure 2.3 A coupled inductor

Fly back Converters or Converters with Isolation

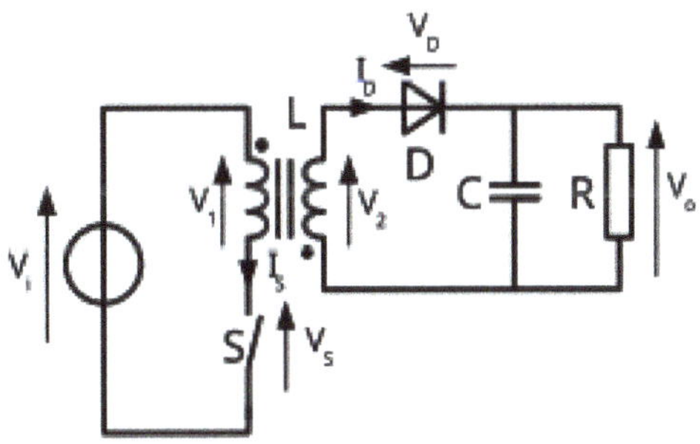

Figure 2.4 Fly back Converter

This converter is suitable for the operation of the power line. The name fly back comes from the analog TV circuitry sweeping the screen and the fly back circuitry to move the screen and bring it back.

This circuit employs a coupled inductor, as shown in Figure 2.5

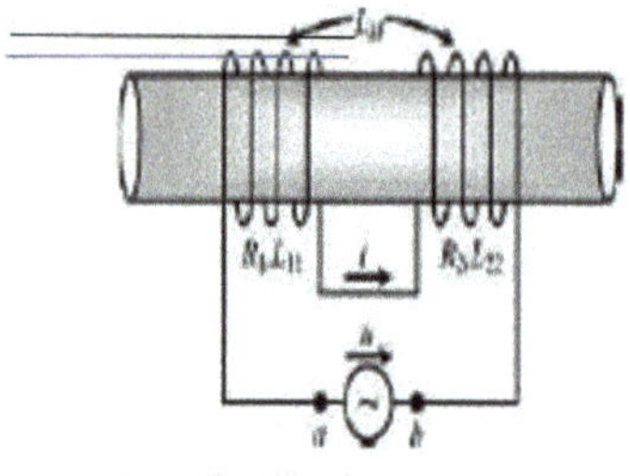

Coupled inductor

n_1 n_2

Figure 2.5 Coupled Inductor

$$\frac{L_1}{L_2} = \left(\frac{n_1}{n_2}\right)^2$$

Coupled inductor looks like a transformer but is not a transformer. In a transformer, a current enters one winding and exits the other winding. There is no energy stored in the magnetic core. In the case of a coupled conductor, one winding can have a current coming in that is stored in the core and released later when required through another inductor. These inductors are connected with each other by the equation above. In an inductor current can be interrupted by a switch that results in very high voltage if the interruption is over a small amount of time. See the currents interrupted by switches S_1 and S_2 flowing through windings with a turns ration of 1:1 and connected to voltage sources V_1 and V_2

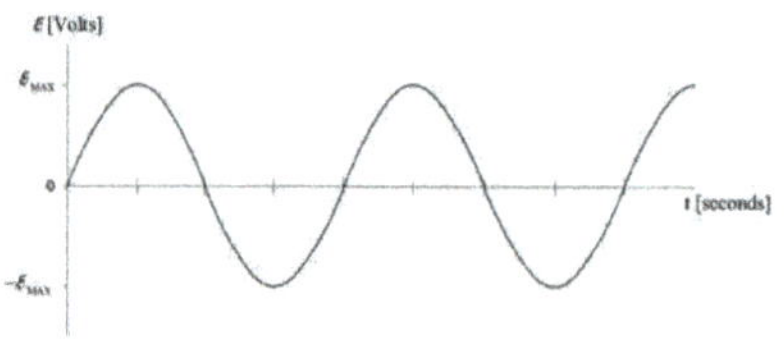

Figure 2.6 Current Winding interruption

This voltage is given by the state space equation as:

$$V = L\frac{dI}{dt} \qquad (2.2)$$

$$V = L\lim_{\Delta t \to 0}\frac{\Delta I}{\Delta t} \qquad (2.3)$$

When $\Delta t \to 0$, this voltage is then very high. The switching is as shown in the above figure. When switch one is on, switch 2 is off, and the change takes place at a single point in the figure. Both the currents flow into the dots of the coils at the instant the switches change their operation, i.e., being on and off. Energy stored in the core gives rise to current in the two coils a shown below:

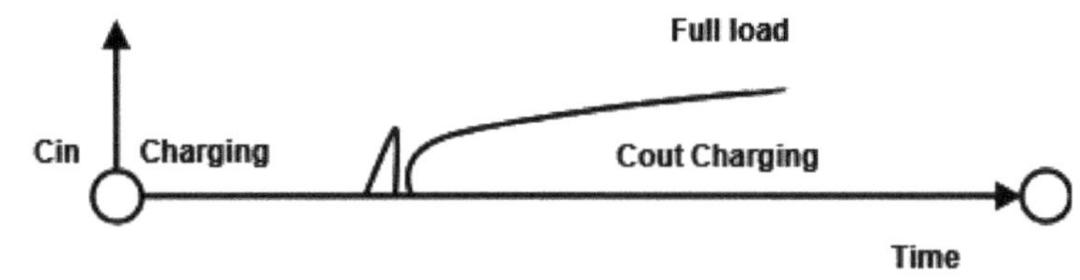

Figure 2.7

At transition, the peak values in the two coils are given by

$$I_{pk1} = I_{pk2} \quad (2.4)$$

And

$$N_1 I_{pk1} = N_2 I_{pk2} \quad (2.5)$$

Thus,

$$\frac{N_1}{N_2} = \frac{I_{pk2}}{I_{pk1}} \quad (2.6)$$

This is because the rate of change of the current is from above equal to V/L, which is a constant, and so the current in the coils rises in the first coil to its peak value and then, at transition, drops to zero as shown, such that.

$$slope\ of\ current\ I_1\ in\ coil\ 1\ is\ \frac{V_1}{L} \quad (2.7)$$

$$slope\ of\ current\ I_2\ in\ coil\ 2\ is\ \frac{V_2}{L} \quad (2.8)$$

Thus, if the ratio is 1:n, then

$$n^2 L = L_2 \quad (2.9)$$

Thus,

$$1. I_{pk1} = I_{pk2}. n \quad (2.10)$$

Which yields,

$$I_{pk1} = \frac{I_{pk2}}{n} \quad (2.11)$$

We now replace the switch with a diode, then in the figure below:

$$V_{10} = nV_1 \quad (2.12)$$

$$V_{20} = \frac{1}{n} V_1 \quad (2.13)$$

And the voltage is negative. Our circuit modification uses one switch and the second as a diode. As shown in the figure below with the turns ratio of 1:n:

Once S_1 is switched on, the current flows through the primary or first coil, and the magnetic core stores some energy. Current flows into the dot of the first coil through switch S_1, and then voltage flows through the second coil also into the dot, which results in the diode conduct and voltage being negative on the V_2 side. We can now proceed to a buck-boost converter.

Inrush current in DC-DC Converters

As shown in the Figure 2.8 we have a DC source in series with an Electromagnetic Interference (EMI) filter in series with the C-DC Converter, which in turn is in series with the load.

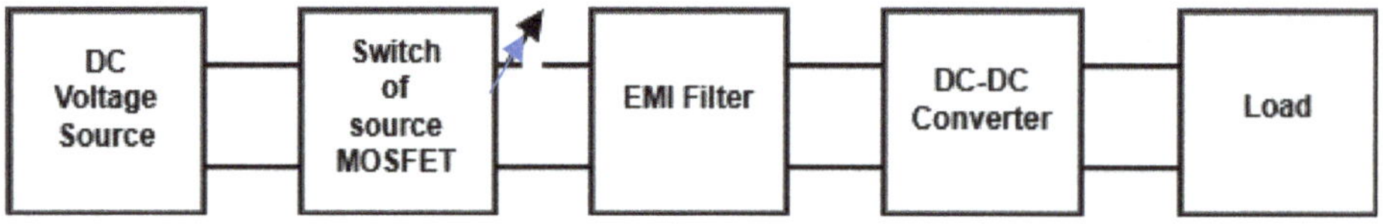

Figure 2.8 Block diagram of a DC-DC Converter.

The initial inrush current is due to the charging of the input capacitor in the DC-DC Converter and the capacitors in the EMI filter are shown in Figure below. The next inrush charging current comes from the charging of the output capacitor in the DC-DC Converter at full load. The fast dv/dt of the input voltage causes the peak circuit to be generated compared to the full load voltage change in the running current at full load. More so because DC-DC converters have a soft start circuit that slowly controls the ramping of the charging of the output capacitor.

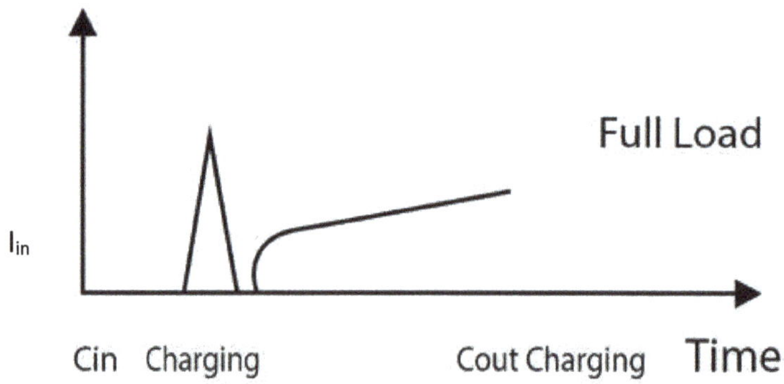

Figure 2.9 Charging and discharging of capacitor of a DC-DC Converter.

Figure 2.9 In-rush current in DC-DC Converter with an EMI filter. The sharp rise in the initial charging current is reduced by using a control circuit, as shown in Figure 2.10 below:

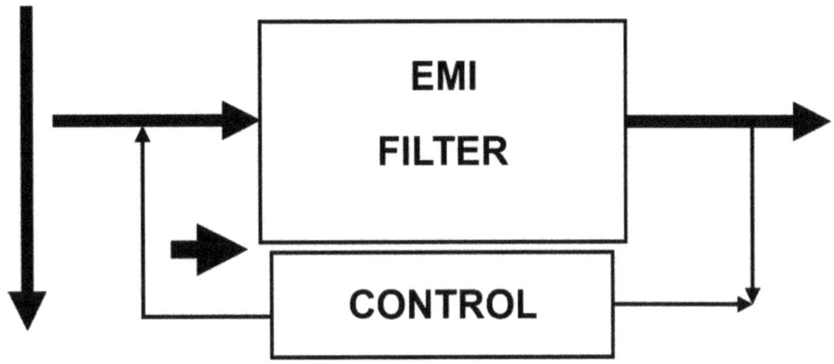

Figure 2.10 EMI filter

The switch is connected in series further with a MOSFET switch, as shown in such that when the switch is shut, the initial charging current is regulated and controlled to a lower level than the previous value and lower than the charging current in the full load current as shown in Figure 2.11 below:

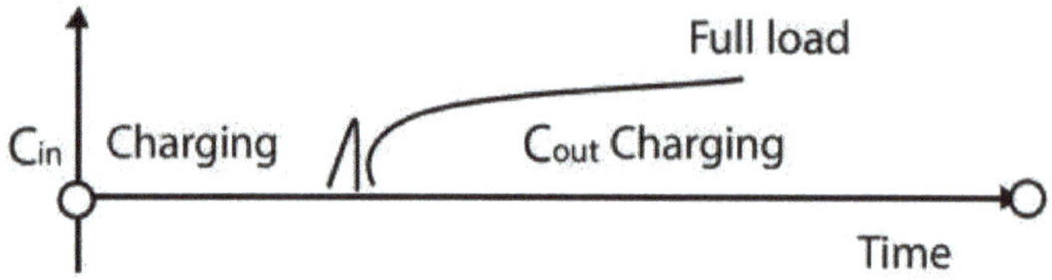

Figure 2.11 Reduction in charging current due to MOSFET switch High Power Boost SEPIC Inverting DC-DC Converters with Fault Protection

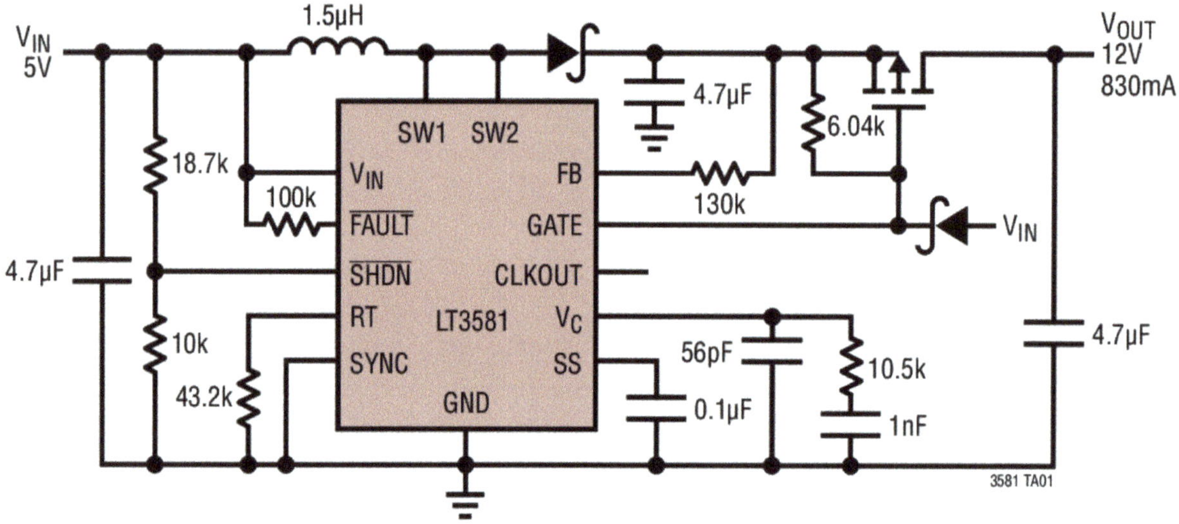

Figure 2.12 SEPIC inverting DC-DC Converter with Fault protection. Courtesy client

This high-power unit generates a 24-volt output from either an 8-volt to 16-volt input. The configuration is shown in figure 2.12 above using LT© 3579/3581 DC-DC converters with fault protection. This circuit can create an output current of 1.1 amps at 24 volts DC. The power switch is integrated inside this chip. For LT 3581, it has an output current of 3.3 amps. Because these are monolithic ICs, higher switching frequencies are possible. The switching frequency can be set as free, running between 2.5 KHz to 25 MHz. Or an external clock can be used to synchronize the ports. The above circuit operates at 1 MHz, allowing the use of a physically small inductor. It has an efficiency of 92%.

Applications include boost converters with output voltages of 40 volts. The boost converters are constructed to be protected from output short circuit conditions.

These types of DC-DC converters are also constructed for SEPIC DC-DC Converter configuration with low input voltage as low as 4 volts.

The LT 3581 can be used to generate an inverting converter where the input is – 12 volts, as shown in the figure below. This circuit can generate 625mamps output current. The third circuit below shows and LT 3581 chip to generate a 5-volt output at 0.9 amps from 3 volts to 16 volts input.

All the above circuits also incorporate innovative fault protection features. With the capacitor between the input side to ground and output side to ground can make the fault current a very large value and difficult to control. This large output fault current can be controlled by adding external components.

Such a circuit has a CMOS, as shown in the figure below.

The PMOS device will turn off thus breaking the path between the input and output path when the fault occurs. The Figure below shows the output current when the fault occurs. The PMOS on and off conditions are for a very short time, after which the device turns on. The same fault current protection circuits can be used to protect against input over voltage conditions and input reverse voltage conditions.

PSpice Example of Boost Converter

Circuit in Figure 2. 13 is a Boost DC-DC Converter circuit drawn Using ORCAD PSpice:

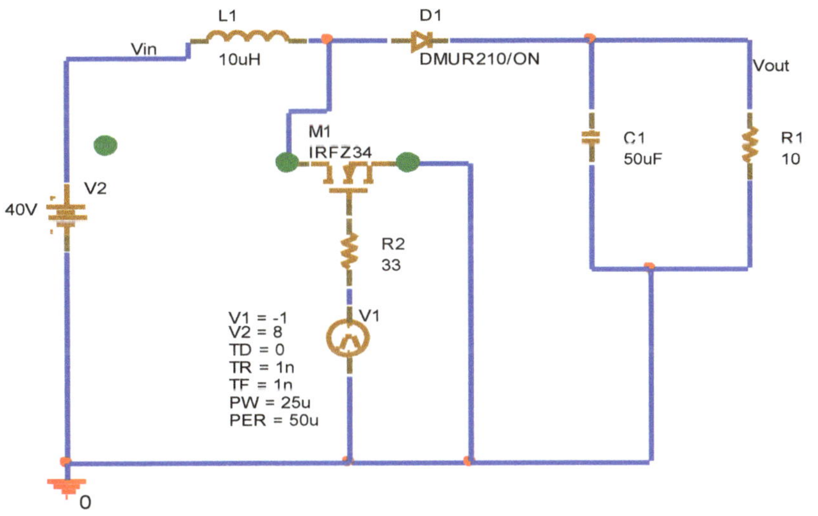

Figure 2.13: Boost DC-DC Converter Circuit drawn in ORCAD PSpice

The circuit shows the different components used in the circuit:

1) 10 micro H-inductor on the front end
2) D1- diode to charge the capacitor across load R1
3) M1 – MOSFET – activated by the pulser
4) V1 – pulser with specifications given on the diagram
5) C1 capacitor across the load, which also provides some filtering action
6) R1 – load resistor
7) V2- input DC voltage, also as V_{in} to the inductor
8) V_{out} at load

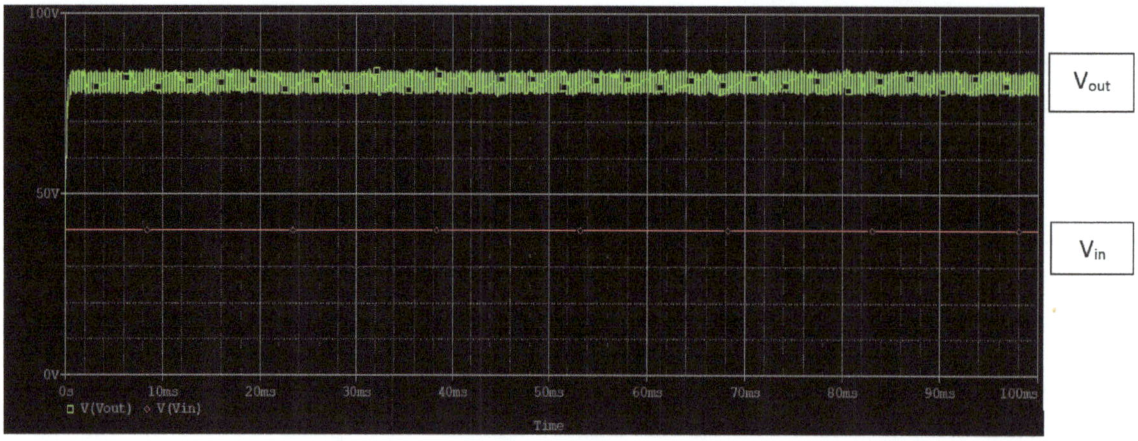

Figure 2.14: Waveform Showing V_{in} and V_{out} for the Boost DC-DC Converter 50-80 DC volts Circuit.

The output is boosted to 80 volts from 40 volts but has a ripple which can be adjusted or reduced in its ripple voltage content with appropriate capacitor values.

Chapter 2 Problems

Problems:

2.1) In your own words, describe the need of DC-DC converters in HVDC transformations.

2.2) Describe the different types of such DC-DC converters.

2.3) What kind of regulation is achieved using such converters?

2.4) Describe a dc-DC converter generally used in automobile industry for electric vehicles (EVS). You can consider any of the current EVs on the market, like Leaf, Tesla, Volt, or REVOLT by GREEN707club at University of Hartford. See Face Book page 'GREEN 707 CLUB 'All the above answers should be your own. You need to do some research for this work or draw upon your undergraduate knowledge in Electronics.

Chapter 3

Buck DC-DC Voltage Converters

In these types of Buck DC-DC Converters, the output voltage V_{out} is at a lower magnitude than the input voltage V_{in}. Some block diagram representations of such Buck converters are shown below in Figure 3.1

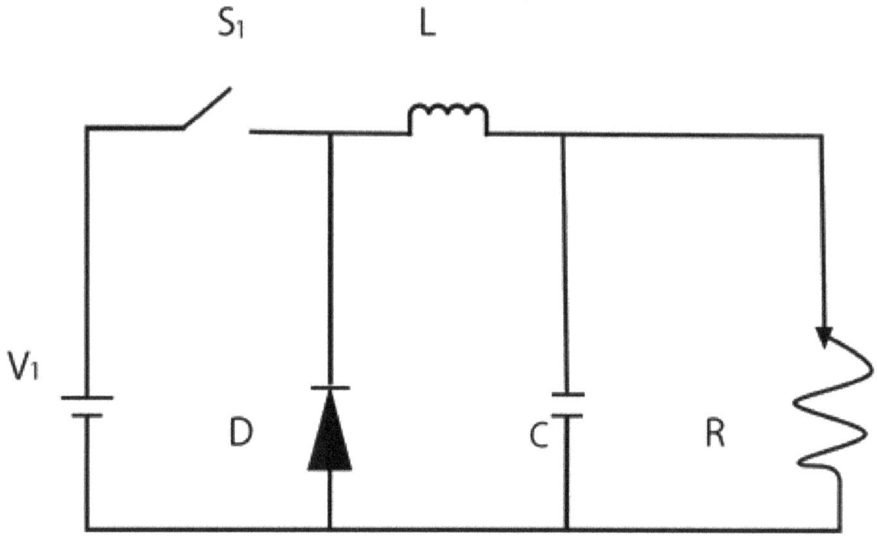

Figure 3.1 Buck DC-DC Converter

A buck converter (step-down converter) is a DC-to-DC power Converter, which steps down voltage (while stepping up current) from its input (supply) to its output (load). It is a class of switched-mode power supply (SMPS) typically containing at least two semiconductors (a diode and a transistor, although modern buck converters frequently replace the diode with a second transistor used for synchronous rectification) and at least one energy storage element, a capacitor, inductor, or the two in combination. To reduce voltage ripple, filters made of capacitors (sometimes in combination with inductors) are normally added to such a converter's output (load-side filter) and input (supply-side filter).

Critical value of inductance of buck-boost converter in boundary conduction mode (BCM)

This is evaluated by looking at the inductance current in the BCM mode, as shown in Figure 3.2 below:

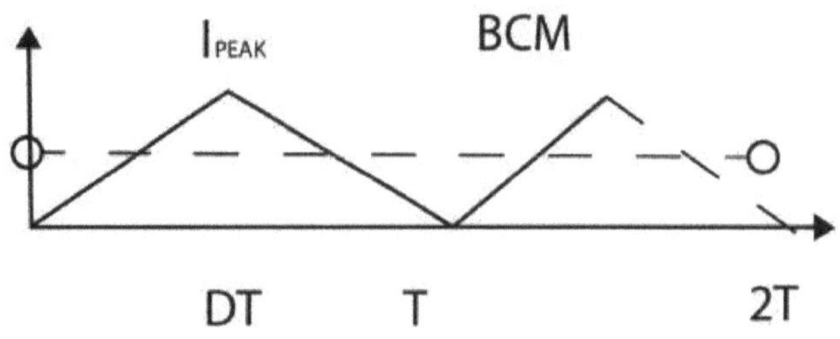

Figure 3.2 Inductance current in a buck-boost converter in the BCM mode of operation.

The inductor current starts at zero and increases during the time DT and hit the peak value i_{peak} and then reduces to zero by the end of the time period T. To calculate this i_{peak} value, we need to know the average inductor current $<i_a>$ shown by the dashed line in the figure above. This is the value of the average current flowing through the inductor which is the sum of all currents flowing through it to can be found by using KCL.

$$< i_L > = < i_i > + i_D = < i_i > + < i_o > + < i_C > \quad (3.1)$$

But i_C is zero in a steady state condition, which results in an average diode current equal to the average output current. Thus,

$$< i_L > = < i_i > + i_D = < i_i > + < i_o > \quad (3.2)$$

And now, the value of the critical inductance can be calculated using the voltage equation as, over the entire time period T

$$V = L\frac{\Delta i}{\Delta T} \quad (3.3)$$

When the switch is on, the input current flows through the inductor, including the average output current, which yields

$$V_i = L_{critical}\frac{i_{peak}}{DT} = L_{critical}\frac{2(<i_i>+<i_o>)}{DT} \quad (3.4)$$

This gives the value of the critical inductance as:

$$L_{crit} = \frac{V_i DT}{2(<i_i>+<i_o>)} \quad (3.5)$$

When the values of the input current and output current are known. If the output load resistance of R is known, then we can make further evaluations to find L_{crit} as follows by finding the average input and output current:

We use the power balance equation; input power is equal to output current:

$$V_i < i_i > = < V_o >< i_o > \quad (3.6)$$

$$< i_i > = < V_o >< i_o > / V_i \quad (3.7)$$

Knowing the average value of output current from Ohm's Law as:

$$< i_o > = \frac{<V_o>}{R} \quad (3.8)$$

Which, when substituted in the previous equation of L_{crit}, yields:

$$L_{crit} = \frac{V_i DT}{2(<V_o><i_o>/V_i + \frac{<V_o>}{R})} \quad (3.9)$$

$$L_{crit} = \frac{V_i DT}{2\left(<V_o><i_o>\frac{1}{V_i} + <i_o>\right)} \quad (3.10)$$

$$L_{crit} = \frac{V_i DTR}{2 < V_o > (1 + \frac{< V_o >}{V_i})} \tag{3.11}$$

Feedback control of DC-DC Buck converter

We will use the standard feedback control block diagram illustrated in the previous generalized representation of the feedback control system. Here we will now model the system as a SISO system and proceed to modelling the plant's different parameters:

1) Model the dynamic states of the system
2) Simplify and linearize the system and stabilize it around the steady state dynamic states, which is around the equilibrium point of the system.
3) Obtain the transfer function for the linearized system.

Here, we will model the system for the dc-dc buck converter. For this, we need to understand the working of the buck converter. The dynamics of the system is controlled by the switch S_1in figure 3.3 below shown again below:

When the switch is on, power from the input is stored in the inductor, and the output capacitor and current flows into the inductor and then passes across the capacitor as the diode is reverse biased (RB).

When the switch is off, the current direction reverses in the capacitor but flows in the same direction as previously initiated and voltage across the capacitor and the current in the inductor starts to discharge through the output resistor.

Although this is the current flow in two different states, here we are concerned with the behavior of the converter with respect to the average current flow in the output resistor I_o.

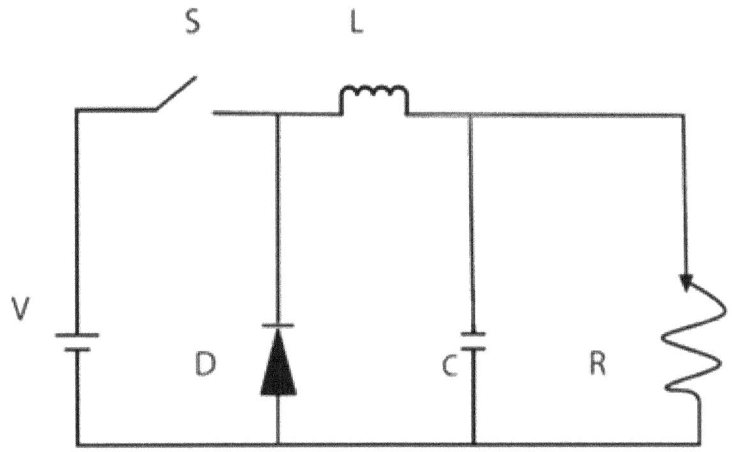

Figure 3.3. DC-DC Buck converter

The control input to the controller is the active duty ratio of the control switch, which is shown in Equation 3.12

$$d = duty\ ratio = \frac{t_{on}}{T} = \frac{t_{on}}{t_{on} + t_{off}} \tag{3.12}$$

Which is the on-time divided by the total period of the switch, as shown above in Figure 3.2.

This is an average value 'd', so we need the performance of the system as an average behavior. This has to be translated into the dynamic states of the system. In the buck converter, the dynamic states are those which exhibit a derivative in the representation of the current in the inductor and voltage across the capacitor. Thus, the inductor current and the capacitor voltage in the buck converter can be written as in Equations (2) and (4), respectively:

Inductor current:

$$\frac{di_L}{dt} = \frac{1}{L}[(V_i - v_c)d + (-v_c)(1-d)] \tag{3.13}$$

In Equation (2), the voltage across the inductor when the switch is on is the first term $(V_i - v_c)d$, and when the switch is off, it is $(-v_c)(1-d)$. Thus, simplifying Equation (3.13) we get:

$$\frac{di_L}{dt} = \frac{1}{L}(V_i)d - \frac{v_C}{L} \tag{3.14}$$

Capacitor voltage:

$$\frac{dv_C}{dt} = \frac{1}{C}\left[\left(i_L - \frac{v_C}{R}\right)d + \left(i_L - \frac{v_C}{R}\right)(1-d)\right] \tag{3.15}$$

Here, we look at the current through the capacitor through each state of the switch. Thus, when the switch is on it is $\left(i_L - \frac{v_C}{R}\right)d$ and when the switch is off the current in the inductor is $\left(i_L - \frac{v_C}{R}\right)(1-d)$; which when simplified, gives:

$$\frac{dv_C}{dt} = \frac{1}{C}\left[(i_L) + \left(-\frac{v_C}{R}\right)\right] \tag{3.16}$$

Equations (3.14) and (3.16) are the two dynamic equations for the buck converter, the inductor current and capacitor voltage. The next step is to linearize these equations and apply the feedback controller. We will use the general form of the PID controller TF. Here, the input to the controller is the error E(s), and the output of the controller, which is the input to the dc-dc converter, is U(s), which is the duty ratio. In the PID controller, we have the integral term, a derivative term, and a proportional term, as shown in Equation (3.16)

PID Controller Transfer Function:

$$G_c(s) = \frac{K_d s^2 + K_p(s) + K_i}{s} \tag{3.16}$$

We will compare this to the transfer function of the closed loop transfer function of the buck converter as shown below in Equation (3.17):

$$\frac{Y(s)}{Y_{ref}(s)} = \frac{\frac{V_i}{LC}G_c(s)}{s^2 + \frac{1}{RC}s + \frac{1}{LC} + \frac{V_i}{LC}G_c(s)} \tag{3.18}$$

Inserting (3.16) in (3.17) yields

$$\frac{Y(s)}{Y_{ref}(s)} = \frac{\frac{V_i}{LC}(\frac{K_d s^2 + K_p(s) + K_i}{s})}{s^2 + \frac{1}{RC}s + \frac{1}{LC} + \frac{V_i}{LC}(\frac{K_d s^2 + K_p(s) + K_i}{s})} \quad (3.18)$$

Simplifying Equation yields Equation (9) as the expression of the closed loop the transfer function of the PID controller of the buck converter yielding:

$$\frac{Y(s)}{Y_{ref}(s)} = \frac{\frac{V_i}{LC}(K_d s^2 + K_p(s) + K_i)}{s^3 + \left(\frac{1}{RC} + \frac{V_i}{LC}K_d\right)s^2 + \left(\frac{1}{LC} + \frac{V_i}{LC}K_p\right)s + \frac{V_i}{LC}K_i} \quad (3.19)$$

The denominator is third-order equation indicating that there will be three poles of the closed-loop system for the buck dc-dc converter. Here, R, L, and C are inherent characteristics of the converter, and these cannot be changed. We can only change the K_d, K_p and K_i parameters of the closed loop PID controller. The performance characteristics of the PID Controller are stability, rise time (t_r), overshoot (OS), and settling time (τ_s). This is possible if the poles are in the left half of the s-plane per control theory. Thus, by choosing proper values of the system parameters, say as follows:

Example 1 on Buck Converter:

In Figure 3.1, following are the components:

R = 10 Ω; L = 100 μH; C = 500 μF; V_i = 12 V

PID Controller

K_p= 0.1; K_i = 1000; K_d = 0.0001

After plugging these values in equation (9), we get this transfer function for the closed loop PID controller of the dc-dc buck converter as in Equation (3.20):

$$\frac{Y(s)}{Y_{ref}(s)} = \frac{2.4x10^4 s^2 + 2.4x10^7 s + 2.4x10^{11}}{s^3 + 2.4x10^4 s^2 + 4.4x10^7 s + 2.4x10^{11}} \quad (3.20)$$

MATLAB results Using work sheet:

```
Command Window
New to MATLAB? See resources for Getting Started.
>> num=[2.4*10^4 4008*10^3 2.4*10^7]

num =

       24000     4008000    24000000

>> den=[1 24200 24008*10^3 2.4*10^7]

den =

           1       24200    24008000    24000000

>> sys=tf(num,den)

sys =

       24000 s^2 + 4.008e06 s + 2.4e07
   -------------------------------------
   s^3 + 24200 s^2 + 2.401e07 s + 2.4e07

Continuous-time transfer function.
```

Figure 3.4: Figure showing written MATLAB code

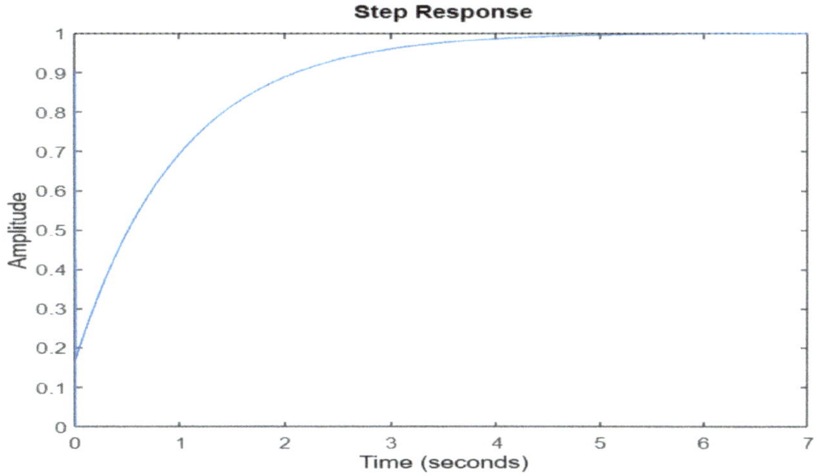

Figure 3.5: Graph depicting the step function of given system.

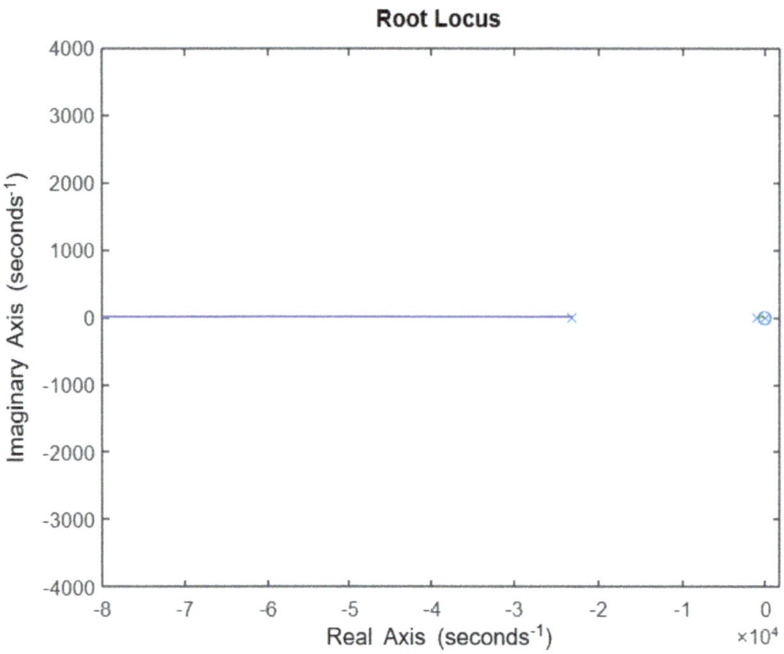

Figure 3.6: Graph showing the root locus of the system

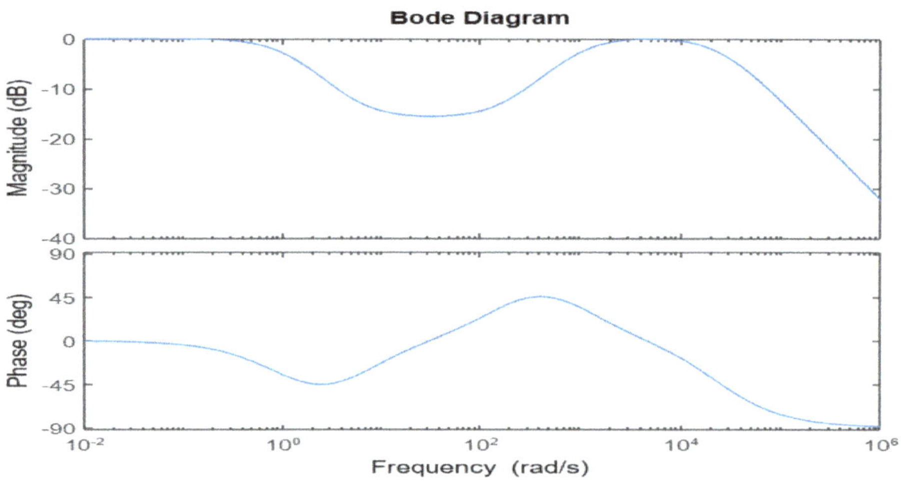

Figure 3.7: Graph depicting the Bode plot of the given system

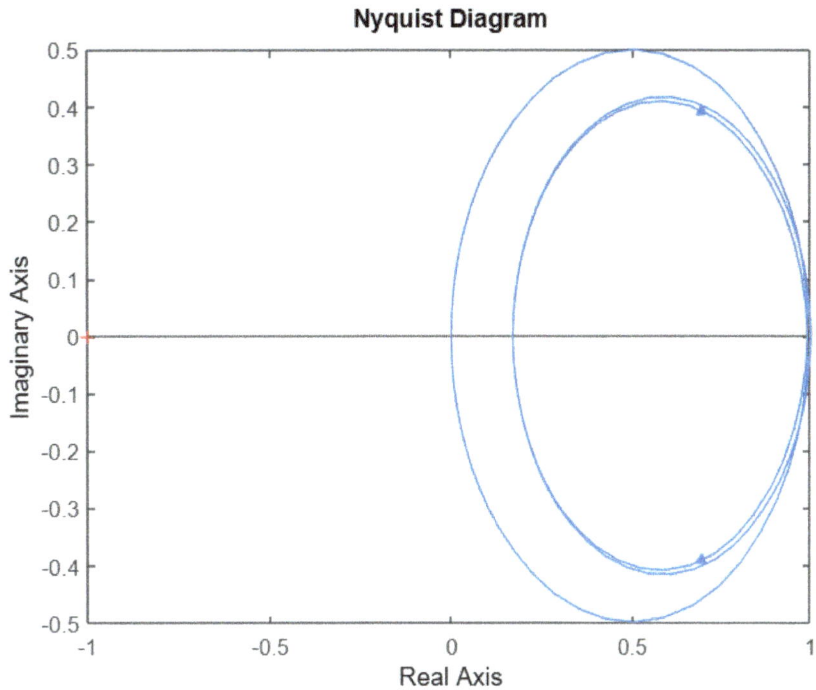

Figure 3.8: Graph showing the Nyquist plot.

Using Matlab to obtain the root locus, we obtain all three poles in the left half s-plane. System is stable. We can obtain the step response of the system with a stable value of the output but with an Over Shoot (OS) of more than 5%. We now change the control values to:

$$K_p = 0.0167$$

$$K_i = 0.1$$

$$K_d = 0.00001$$

To obtain a new transfer function as in Equation (3.21)

$$\frac{Y(s)}{Y_{ref}(s)} = \frac{2400s^2 + 4.0x10^6 s + 2.4x10^7}{s^3 + 2600s^2 + 2.4x10^7 s + 2.4x10^7} \tag{3.21}$$

MATLAB results Using work sheet:

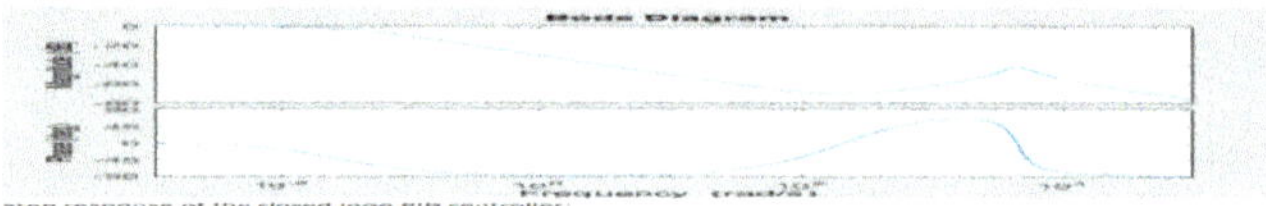

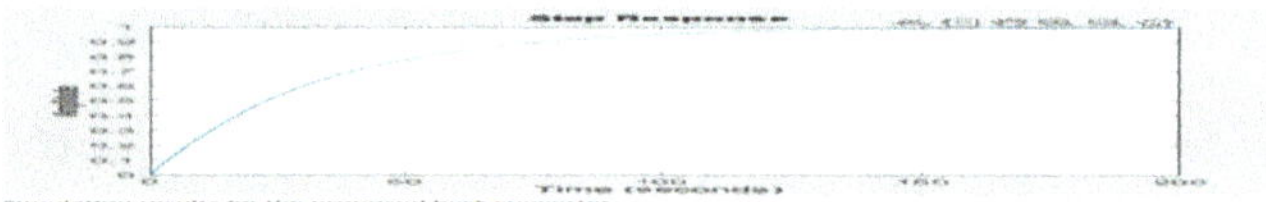

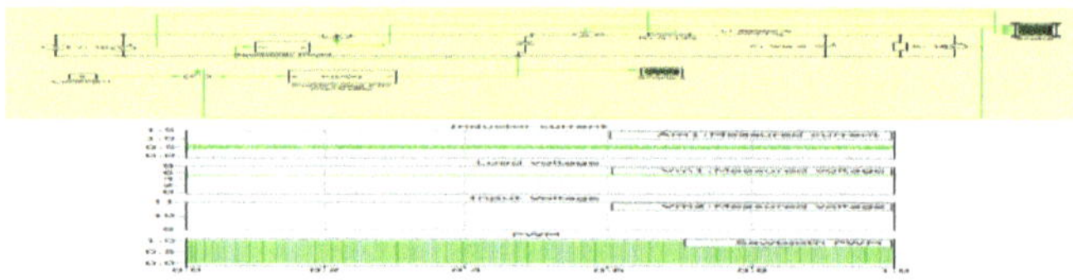

Using MATLAB, we can obtain the root locus of the closed loop transfer function of the PID Controller of the dc-dc buck converter and the step response as well, Root locus shows all the poles are in the left half plane. However, the pole on the real axis has now moved close to the origin but is still in the left half plane, and the complementary poles have moved away from the origin. The step response is very slow, but the Over Shoot (OS) is no longer present. The output of the converter remains constant at 5 volts.

Control System in DC

Feedback Control

It has three basic elements

1) The plant or System

2) The control Objective

3) Controller

4) The plant or the System – in a more general case s-system has one or more inputs (u_p) and one or more outputs (y_i) as shown in Figure 3.9. The System has states denoted by x_i; i=1.2…..n. These states and the inputs generate one or more outputs.

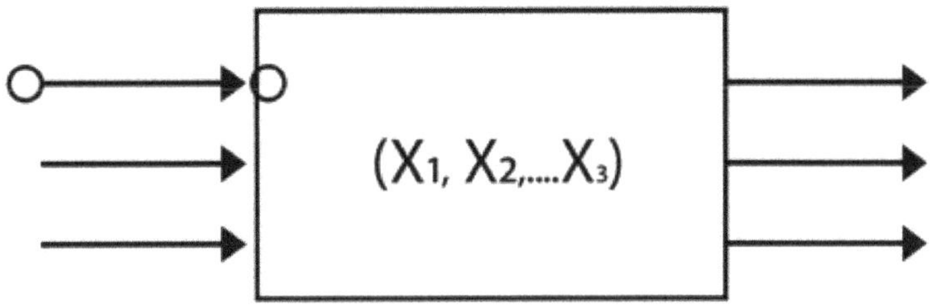

Figure 3.9 System diagram of a plant along with inputs and outputs

In the boost. Buck and the buck-boost converters the states are the inductor current and output capacitor voltage. For the input the switch input is the input duty ratio. For the output, the output current and voltage, along with the input current, are outputs. Besides, it is important to remember that the input and output of the converter are not the same as system input and output power. For control systems can be simple or complicated. For feedback control, we need to simplify and linearize the system definition in order, to work with them very easily. For feedback control, it is easier to work with single input single output system, or SISO system. So in Figure 3.9, we will have duty cycle of the switch 'd' as input, and for dc-dc converters, we have a targeted output. Then linearize the system as shown in Figure 2 and simplify such that the SISO system for a dc-dc converter has 'd' as the input and the output capacitor voltage as the output of the simple and linearized SISO system.

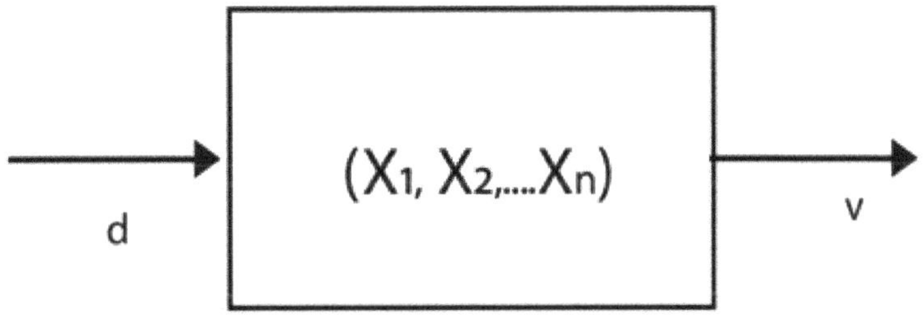

Figure 3.10. SISO system of dc-dc converter with switch duty cycle 'd' as the input and output capacitor voltage as the output

1) The Control Objective of a dc-dc converter

The targeted output for a dc-dc converter is generally the constant DC output voltage, but it could also be output current or output inductor current. The next step is to choose the target value for this output. Say this 5volts as in the case of an iPhone. The general diagram of the feedback control as shown in Figure 311.

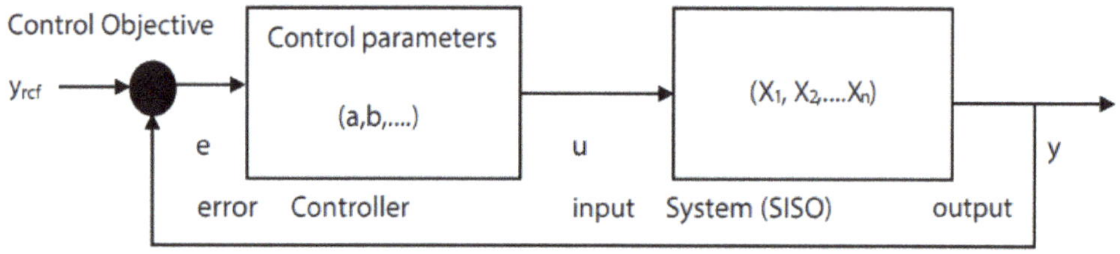

Figure 3.10. System of dc-dc converter with a feedback control.

For the feedback control, we take the actual output (y) of the system and subtract it from the reference output voltage (y_{ref}) to get the error.

2) The Controller - This error 'e' is fed into the controller, and the controller should decide the correct input to the system that will cause the system output to reach the control objective. In our example, this is 5 volts. The controller, thus, is a key component in the design of an efficient feedback control system with high performance. Atypical standard feedback controller is proportional-integral-derivative (PID) controller. There are many different types of controllers.

We will add these controllers in the sequel.

PSpice Example of Buck Converter

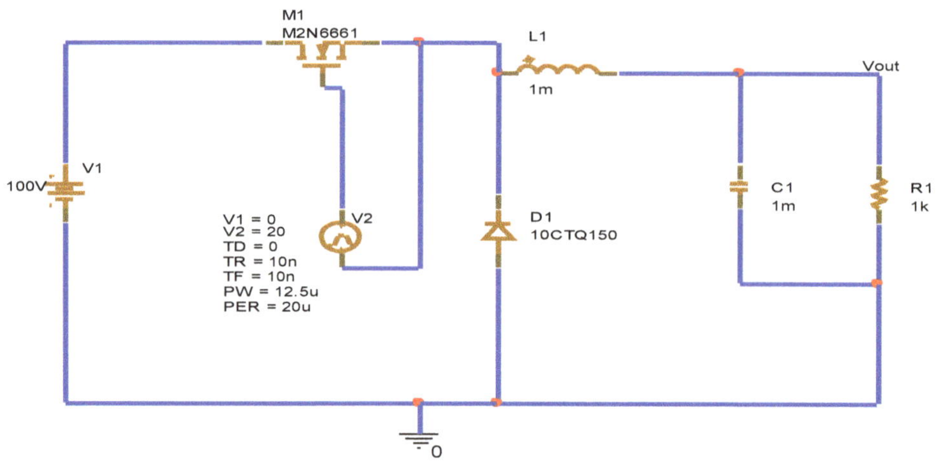

Figure 3.11: Buck DC-DC circuit drawn in OrCAD PSpice

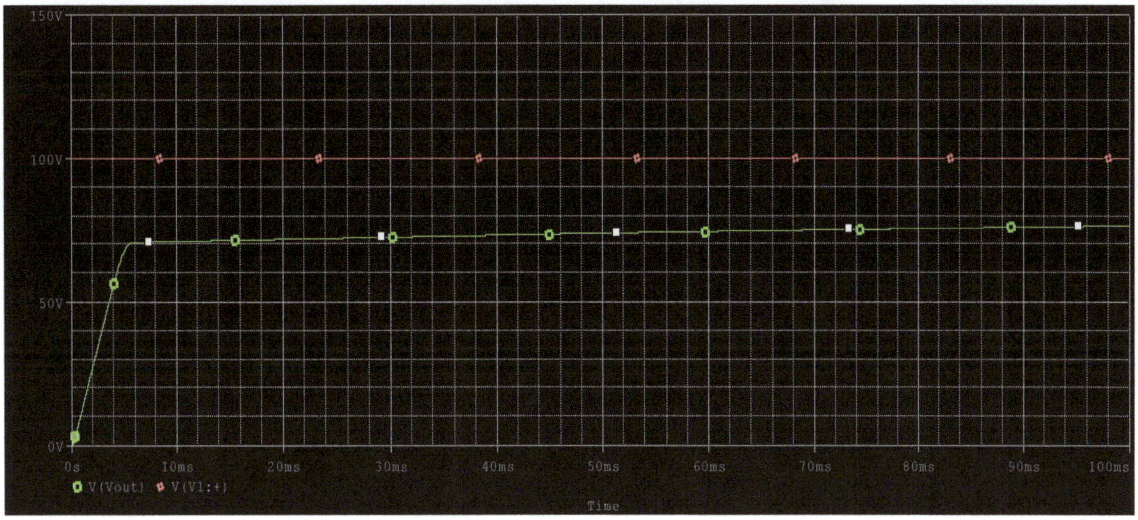

Figure 3.12: Image showing the waveforms of V_{in} and V_{out}

Chapter 3 Problems

Problem 1:

Design the PID controller for Buck DC-DC Converter per the values of Kp, Kd, and Ki given

Below:

K_p= 0.02

K_i = 0.15

K_d = 0.000015

Please provide:

1) Step response
2) Root locus
3) Bode plot
4) Nyquist plot

Using MATLAB for the closed loop Transfer Function.

Chapter 4

Boost Converters

Boost Converter

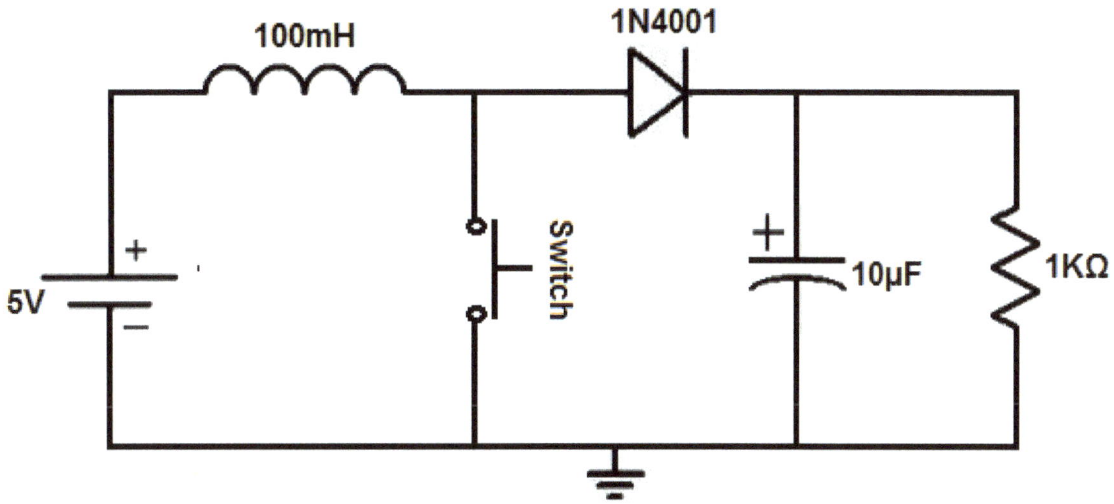

A boost converter (step-up converter) is a DC-to-DC power converter that steps up voltage (while stepping down current) from its input (supply) to its output (load). It is a class of switched-mode power supply (SMPS) containing at least two semiconductors (a diodc and a transistor) and at least one energy storage element: a capacitor, inductor, or the two in combination. To reduce voltage ripple, filters made of capacitors (sometimes in combination with inductors) are normally added to such a converter's output (load-side filter) and input (supply-side filter).

A DC-to-DC boost converter circuit is a circuit that can convert a DC voltage into a larger DC voltage.

So, for example, you may be able to convert a 5V DC voltage into 30V.

A DC-to-DC converter works on the principle of an inductor primarily and a capacitor. When fed DC power, the inductor acts as an energy storage device for current. As long as DC power is supplied to it, it and builds up current through the coils as well as a magnetic field around itself. When the DC power is turned off, the magnetic field collapses, and all the current that the inductor charged up gets dumped across to the capacitor. As the DC power gets turned on and off in the circuit, which we will do through a pushbutton switch, the inductor stores up more current and dumps it across the capacitor each time. This is why the voltage across the capacitor increases each time we do this until it reaches its peak.

The amount of voltage that the output will increase depends upon a number of factors, including the value of the inductor, the maximum voltage storage of the capacitor, the switching speed, and the DC input voltage of the circuit.

We won't go too in depth mathematically about each relationship but will show how this variable relate to the amount of voltage increase there will be.

In this circuit, we will use a manual pushbutton to switch the DC value on and off.

In a more advanced, real-life circuit, you would use a transistor instead of a manual switch, as this gives more precision and control over the circuit. Unlike manual control, a transistor can switch on and off much quicker, in the order of microseconds, and give much more precision. A microcontroller would control the transistor, and it could be coded to turn the transistor on and off at set intervals. Under such control, the output voltage would be able to calculated precisely, and the code could be written so that these bursts of DC voltage occurred at desired times.

However, for this circuit, we are simply demonstrating how a DC-to-DC boost converter works and using a Pushbutton will do the job just fine for this circuit.

Components Needed

- 1K ohms resistor; Diode; Pushbutton switch; 100mH inductor; 10 micro F electrolytic capacitor So, for this circuit, we will use basic discrete components.

This includes a resistor, an inductor, an electrolytic capacitor, and a diode.

In this circuit, we will use a 100mH inductor. Increasing the inductance allows for greater voltage output, because the greater the inductance, the more energy it can store. So, if you swap out the inductor for a lower value, it will be able to produce less output voltage. If you swap out the inductor for a higher value, it will be able to put out a higher voltage. So, you can play around with inductance values to see how it affects the output voltage. If you do use a higher inductance, it will resist current flow through it since it's a greater value, so the charge-up process may take longer, but in the end, it produces a higher voltage output.

The electrolytic capacitor should have a voltage rating of about 50V or higher. The higher the rating, the better, because the more voltage it will be able to handle. You do not want the voltage across the capacitor to exceed its voltage rating, or it may explode. Being that the boost converter circuit can produce tremendously high voltage, you want as high a rating as you can get, at least definitely well over the 25V rating, which is common with capacitors. You want at least 50V.

You also want the capacitor to be of a fairly high value. This is why we're not using Pico or nano-farads for the capacitor but at least 1 microfarad. Since the inductor is putting out large amounts of current, we want the capacitor to have enough capacitance to store this charge.

The diode can really be any diode. We use a 1N4001 diode just because it's easy to find and readily available, even though it isn't the best choice because it uses about 0.7V of the voltage drop across it.

Take into account, though, that whatever diode you use will produce a voltage drop across the diode.

Some diodes use more voltage drop than others, so ideally, you want to use a diode that has a low voltage drop for as little voltage consumption as possible. This would be for real-life situations where you want as much efficiency as possible in terms of possible voltage output. However, for this circuit, you may not care about the less than 1V lost across the diode. But, just so that you know, the diode will consume some of the output voltage current, making it lower than it would be if you used a less power-consuming diode.

The diode is used in this circuit to provide resistance so that when the pushbutton is pressed, all the current flows from the inductor to the ground, creating a short circuit. This is so that the inductor alone gets charged up at this time. When the pushbutton is released, an open circuit is formed at the pushbutton junction, so that all the current then gets discharged through the diode and to the capacitor.

The resistor isn't really necessary for this circuit. It simply acts as the load for the circuit. If you have an actual load that you want powered on, you could replace the resistor for it. Being that the resistor is in parallel to the capacitor, the voltage across the resistor is the same as the voltage across the capacitor, because voltage in parallel is equal.

Boost Converter Circuit

The boost converter circuit is shown below in Figure 4.1.

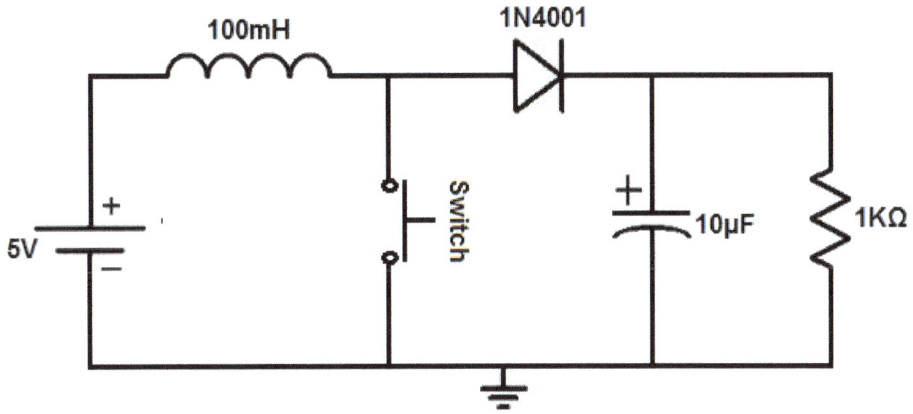

Figure 4.1 DC-DC Boost converter

So we have 5V as the DC power powering this circuit.

If you were to measure the voltage now across the resistor, it would be about 4.3V, less than the 5V supplied because of the voltage drop across the diode.

Now, to increase the output voltage, we press the pushbutton several times.

You don't want to press the pushbutton and hold it long at all, less than a second. This is because when you press the pushbutton, all of the current flows across the inductor and then down to the ground, basically creating a short circuit for the inductor. Inductors do not like large amounts of current going through it alone for long. It could fry the inductor. This is why you want to press the pushbutton and release it, press it, and release it. This allows for an even distribution of current buildup in the inductor and release to the capacitor. It charges up in increments, not all at once.

So, again, you press the pushbutton several times, charging up the capacitor gradually. Since a capacitor charges logarithmically, you will see a logarithmic increase in voltage over time, such as the waveform shown above. As you continue pressing the pushbutton, you will see that it begins to peak out a little above 20V.

How this circuit works is that when you press down on the pushbutton, current flow is created through the inductor down to the ground via the pushbutton. At this point, all the current flows through the inductor to the ground, the lowest path of resistance. No current flows across the diode. This charge up the inductor.

The inductor acts as an energy storage device for the current. The greater the inductance, the more energy it is able to store. So, the inductor charges up.

Once the pushbutton is released, the junction at the pushbutton becomes an open circuit, so no more current can flow through this junction. The inductor discharges all the current built up on it and dumps it through the diode and to the capacitor, charging up the capacitor. This is when the voltage across the capacitor starts building up. As we keep pressing and releasing the pushbutton, the charge continues to increase, building up the voltage further. This happens until the voltage peaks off.

The voltage peak is determined by a number of factors, including the input voltage, the inductance of the inductor, the type of diode being used, the capacitance of the capacitor, and the voltage rating of the capacitor. To increase the voltage output, we would have to increase each factor listed, with the exception of the voltage drop across the diode, which we would want to be as low as possible.

Regarding the input that we want to the inductor, in an actual real-life circuit, we want pulses, such as a pulse width modulated signal, to go to the inductor. This is a square wave with a certain duty cycle that turns on and off, on and off, at determined intervals. We never want to feed the inductor a constant current, because this would make it fry. So, if you were to build this circuit for real-life, industrial use, we would not be using manual control but set up a circuit to deliver pulse width modulated signals to it. This would allow the inductor to charge up and discharge, charge up and discharge, without the inductor receiving so much current that it would be destroyed. So, this illustrates the point why we want to press it and release the pushbutton within a second or less to mimic a square waveform. Again, in real life, this circuit would not be controlled manually by a pushbutton but with a transistor and not with a constant DC power supply but a pulse width modulated signal. But this circuit is just to demonstrate manually how a DC-to-DC boost converter circuit operates.

There are several variations that we can do to this circuit to produce drastic differences in terms of the voltage output.

The biggest change we can make to the circuit to produce drastically different results at the output is to replace the inductor with another component.

The inductor can be replaced with the coil of a relay. This will up the voltage a little over the inductor.

The inductor can also be replaced with a step-up transformer. If using the primary coil, the voltage can be boosted as high as over 60VDC. If using the secondary coil of the transformer, the voltage can be boosted over 120VDC. The secondary coil is a larger coil, which has a greater inductance, allowing for increased voltage at the output. Know that when you are using a step-up transformer instead of the inductor, the capacitor should have a voltage rating of 200V or greater. If less than this, the capacitor can be destroyed, as these are very high voltages. And this is how a DC-to-DC boost converter circuit can be built.

The Output Capacitance value for a Boost Converter

As the inductor is in the input position, as shown in the boost converter circuit in Figure 1, the value of the capacitance does not depend on it but rather on the output voltage. Besides, this value of the capacitor voltage is to be upheld within the incremental voltage ripple that is given within which limit the changes need to take place. Thus, in the switch-on condition, the inductor current flows as shown in the circuit, charging the inductor and accordingly charging the capacitor with the current flowing into the capacitor (i.e., in the negative direction. Thus, the average value of the current is shown in Equation (1) as:

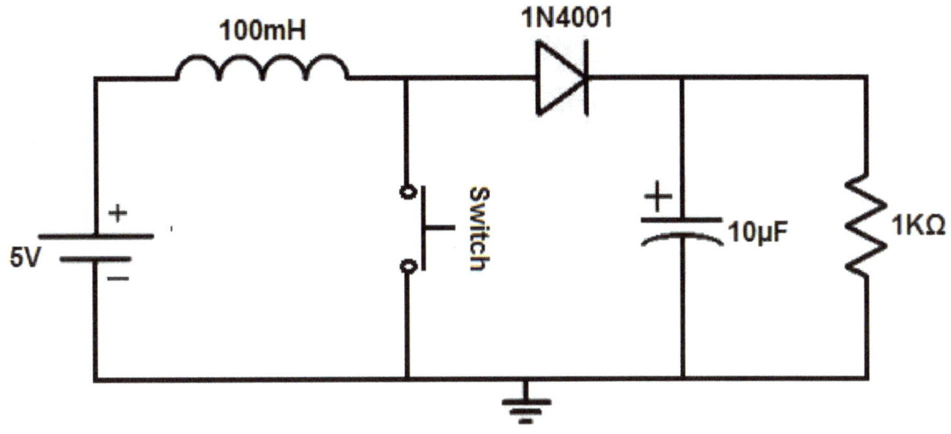

Figure 4.2. Current in the inductor when the switch is on.

$$i_c = C\frac{dV}{dt} = C\frac{\Delta V}{\Delta t} \qquad (4.1)$$

which yields

$$\mathrm{C} = i_c\frac{\Delta t}{\Delta V} =< i_o > \frac{DT}{\Delta V_o} = \frac{<V_o>DT}{R\Delta V_o} \qquad (4.2)$$

The current direction when the switch is on is shown in the Figure 1. Thus, current flows into the capacitor to drive the voltage across the load resistor. We assume that the ΔV_o is a given value. This is the target ripple value of the average output voltage above the average output voltage value.

This value of the output charge current is actually, in reality, a negative value as shown in Figure 4.3, the current plot, but as ΔV is negative, the sign does not show in Equation (1). This value, when the switch is off, has to balance out over the rest of the duty period (T-ΔT).

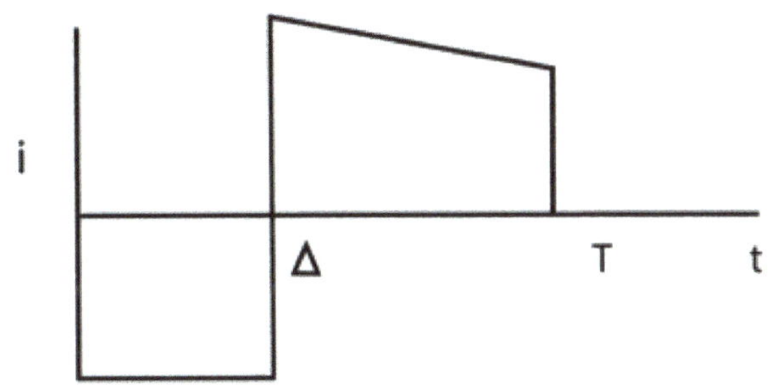

Figure 4.3 Capacitor current for a boost converter

There is no charge hat here, and it is a straight forward calculation independent of the value of the input inductor. Besides, the output part is just a DC RC circuit, and the charge is going to decrease, so to show the real situation, the capacitor current decreases very little, as shown by the dotted line, but the average value stays as the bold line for calculation purposes.

Chapter 4 Problems

Problem P4.1:

Design the PID controller for Boost DC-DC Converter per the values of Kp, Kd, and Ki given Below:

K_p= 0.02

K_i = 0.15

K_d = 0.000015

Please provide

1) Step response
2) Root locus
3) Bode plot
4) Nyquist plot

Chapter 5

Buck-Boost Converters

Buck–Boost Converter

The buck-boost converter is a type of DC-to-DC converter that has an output voltage magnitude that is either greater than or less than the input voltage magnitude. It is equivalent to a fly back converter using a single inductor instead of a transformer.

The **buck-boost converter** is a type of DC-DC converter that has an output voltage magnitude that is either greater than or less than the input voltage magnitude. It is equivalent to a fly back converter using a single inductor instead of a transformer.

Two different topologies are called *buck-boost converter*. Both of them can produce a range of output voltages, ranging from much larger (in absolute magnitude) than the input voltage down to almost zero.

The inverting topology

The output voltage is of the opposite polarity than the input. This is a switched mode power supply with a similar circuit topology to the boost converter and the buck converter. The output voltage is adjustable based on the duty cycle of the switching transistor. One possible drawback of this converter is that the switch does not have a terminal at ground; this complicates the driving circuitry. However, this drawback is of no consequence if the power supply is isolated from the load circuit (if, for example, the supply is a battery) because the supply and diode polarity can simply be reversed. When they can be reversed, the switch can be on either the ground side or the supply side.

A buck (step down) converter combined with a boost (step-up) converter

The output voltage is typically of the same polarity as the input and can be lower or higher than the input. Such a non-inverting buck-boost converter may use a single inductor, which is used for both the buck inductor mode and the boost inductor mode, using switches instead of diodes,[2][3] sometimes called a **"four-switch buck-boost converter"**,[4] it may use multiple inductors but only a single switch, as in the SEPIC and Cuk topologies.

Principle of operation of the inverting topology:

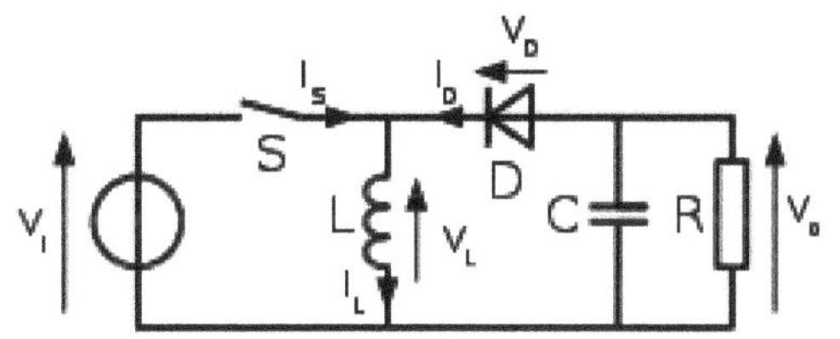

Fig. 5.1: Schematic of a buck–boost converter.

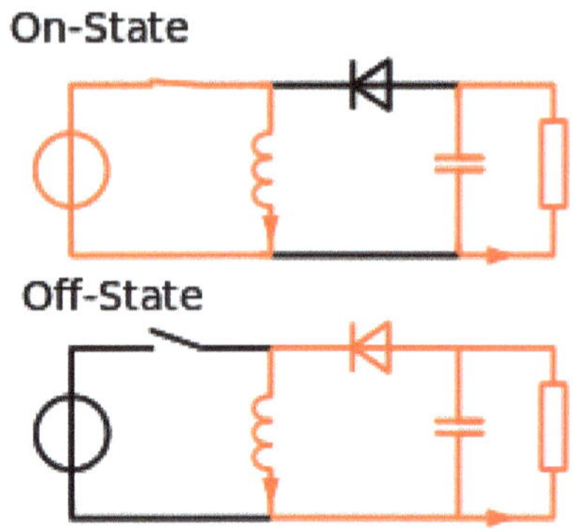

Fig. 5.2: The two operating states of a buck-boost converter: When the switch is turned on, the input voltage source supplies current to the inductor, and the capacitor supplies current to the resistor (output load). When the switch is opened, the inductor supplies current to the load via diode D.

The basic principle of the inverting buck-boost converter is fairly simple (see figure 5.2):

While in the on state, the input voltage source is directly connected to the inductor (L). This results in accumulating energy in L. In this stage, the capacitor supplies energy to the output load. While in the Off-state, the inductor is connected to the output load and capacitor, so energy is transferred from L to C and R.

Compared to the buck and boost converters, the characteristics of the inverting buck-boost converter are mainly:

- Polarity of the output voltage is opposite to that of the input;
- The output voltage can vary continuously from 0 to (for an ideal converter). The output voltage ranges for a buck and a boost converter are respectively 0 and to 100 volts.

Conceptual overview:

Like the buck and boost converters, the operation of the buck-boost is best understood in terms of the inductor's "reluctance" to allow rapid change in current. From the initial state in which nothing is charged and the switch is open, the current through the inductor is zero.

When the switch is first closed, the blocking diode prevents current from flowing into the right-hand side of the circuit, so it must all flow through the inductor. However, since the inductor doesn't allow rapid current change, it will initially keep the current low by dropping most of the voltage provided by the source. Over time, the inductor will allow the current to slowly increase by decreasing its voltage drop. Also, during this time, the inductor will store energy in the form of a magnetic field.

Continuous Conduction mode:

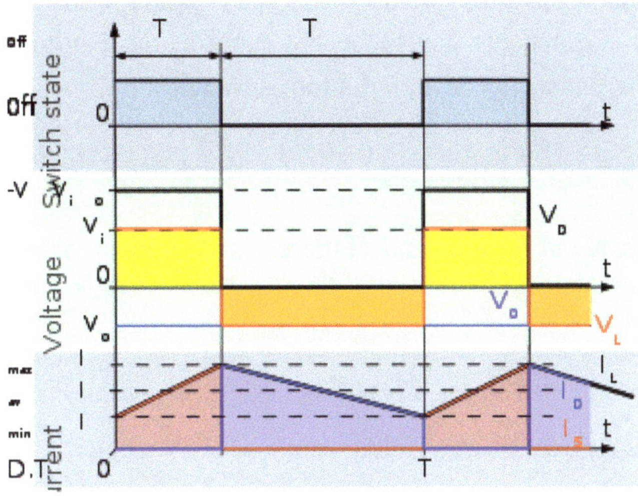

Fig 5.3: Waveforms of current and voltage in a buck-boost converter operating in continuous conduction mode.

If the current through the inductor L never falls to zero during a commutation cycle, the converter is said to operate in continuous mode. The current and voltage waveforms in an ideal converter can be seen in Figure 5.3.

From zero to T, the converter is in On-State, so the switch S is closed. The rate of change in the inductor current (I_L) is therefore given by:

$$\frac{dI_l}{dt} = \frac{V_i}{L} \qquad (5.1)$$

At the end of the On-state, the increase of I_L is therefore:

$$\Delta I_{Lon} = \int_0^{DT} dI_L = \int_0^{DT} \frac{V_i}{L}\, dt = \frac{V_i DT}{L} \qquad (5.2)$$

D is the duty cycle. It represents the fraction of the commutation period T during which the switch is On. Therefore, D ranges between 0 (S is never on) and 1 (S is always on).

During the off-state, the switch S is open, so the inductor current flows through the load. If we assume zero voltage drop in the diode and a capacitor large enough for its voltage to remain constant, the evolution of I_L results in as shown below:

$$\frac{dI_l}{dt} = \frac{V_o}{L} \qquad (5.3)$$

Therefore, the variation of I_L during the off-period is:

As we consider that the converter operates in steady-state conditions, the amount of energy stored in each of its components has to be the same at the beginning and at the end of a commutation cycle. As the energy in an inductor is given by:

$$\Delta I_{Loff} = \int_0^{(1-D)T} dI_L = \int_0^{(1-D)T} \frac{V_o}{L}\, dt = \frac{V_o(1-D)T}{L} \qquad (5.4)$$

It is obvious that the value of I_L at the end of the Off state must be the same as the value of I_L at the beginning of the On-state, i.e., the sum of the variations of I_L during the on and the off states must be zero:

$$\Delta I_{Lon} + \Delta I_{Loff} = 0 \qquad (5.5)$$

Substituting and by their expressions yields results as shown below.

This can be written as:

$$\frac{V_o}{V_i} = -\frac{D}{1-D} \qquad (5.6)$$

This in return yields that:

$$D = \frac{V_o}{V_o - V_i} \qquad (5.7)$$

From the above expression, it can be seen that the polarity of the output voltage is always negative (because the duty cycle goes from 0 to 1) and that its absolute value increases with D, theoretically up to minus infinity when D approaches 1. Apart from the polarity, this converter is either step-up (a boost converter) or step-down (a buck converter). Thus, it is named buck-boost converter.

Discontinuous mode:

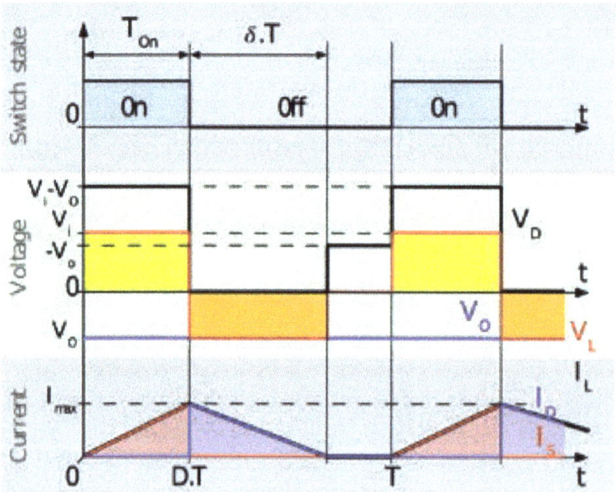

Fig 5.4: Waveforms of current and voltage in a buck-boost converter operating in discontinuous mode.

In some cases, the amount of energy required by the load is small enough to be transferred in a time smaller than the whole commutation period. In this case, the current through the inductor falls to zero during part of the period. The only difference in the principle described above is that the inductor is completely discharged at the end of the commutation cycle (see waveforms in Figure 5.4).

Although slight, the difference has a strong effect on the output voltage equation. It can be calculated as follows:

Because the inductor current at the beginning of the cycle is zero, its maximum value I_{Lmax} (at t = DT) is

$$I_{Lmax} = \frac{V_i DT}{L} \qquad (5.8)$$

During the off-period, I_L falls to zero after δT:

$$I_{Lmax} + \frac{V_o \delta T}{L} = 0 \qquad (5.9)$$

Using the two previous equations, δ is:

$$\delta = -\frac{V_i D}{V_o} \qquad (5.10)$$

The load current is equal to the average diode current (I_D). As can be seen on figure 5.4, the diode current is equal to the inductor current during the off-state. Therefore, the output current can be written as:

$$I_o = \bar{I}_D = \frac{I_{Lmax}}{D}\delta \qquad (5.11)$$

Replacing I_{Lmax} and δ by their respective expressions yields:

$$I_o = -\frac{V_i DT}{2L}\frac{V_i D}{V_o} = -\frac{{V_i}^2 D^2 T}{2LV_o} \qquad (5.12)$$

Therefore, the output voltage gain can be written as:

$$\frac{V_o}{V_i} = -\frac{V_i D^2 T}{2LI_o} \qquad (5.13)$$

Compared to the expression of the output voltage gain for the continuous mode, this expression is much more complicated. Furthermore, in discontinuous operation, the output voltage not only depends on the duty cycle but also on the inductor value, the input voltage and the output current.

Limit between continuous and discontinuous modes:

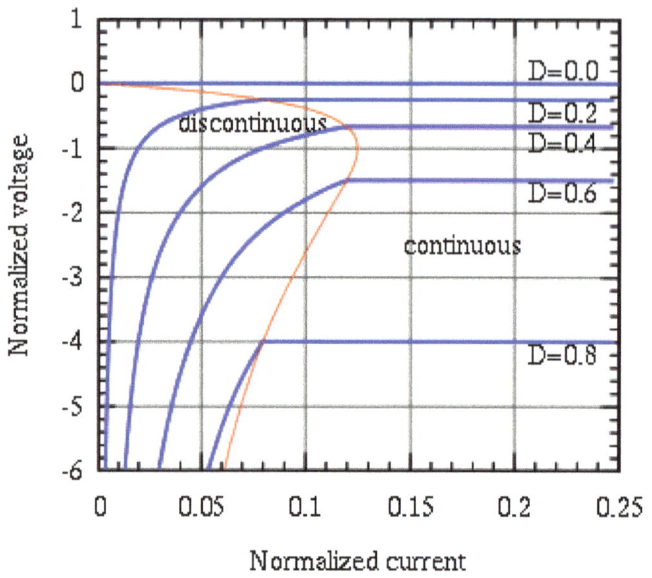

Fig 5.5: Evolution of the normalized output voltage with the normalized output current in a buck-boost converter.

As told at the beginning of this section, the converter operates in discontinuous mode when low current is drawn by the load, and in continuous mode at higher load current levels. The limit between discontinuous and continuous modes is reached when the inductor current falls to zero exactly at the end of the commutation cycle. With the notations of figure 5.4, this corresponds to:

$$DT + \delta T = T \qquad (5.14)$$

$$D + \delta = 1 \qquad (5.15)$$

In this case, the output current I_{Olim} (output current at the limit between continuous and discontinuous modes) is given by:

$$I_{Olim} = \bar{I}_D = \frac{I_{Lmax}}{2}(1 - D) \qquad (5.16)$$

Replacing I_{Lmax} by the expression given in the *discontinuous mode* section yields:

$$I_{Olim} = \frac{V_i DT}{2L}(1 - D) \qquad (5.17)$$

As is the current at the limit between continuous and discontinuous modes of operations, it satisfies the expressions of both modes. Therefore, using the expression of the output voltage in continuous mode, the previous expression can be written as:

Let's now introduce two more notations:

The normalized voltage, defined by $|V_o| = (V_o)/(V_i)$. It corresponds to the gain in voltage of the converter; the normalized current, defined by $|I_O|=(L/TV_i)I_O$. The term $(TV_i)/L$ is equal to the maximum increase of the inductor current during a cycle; i.e., the increase of the inductor current with a duty cycle D=1. So, in steady state operation of the converter, this means that $|I_0|$ equals 0 for no output current, and 1 for the maximum current the converter can deliver.

Using these notations, we have:

1)in continuous mode, $|V_0|=-(D)/(1-D)$; (5.18)
2)in discontinuous mode, $|V_0|=-(D^2)/(2|I_0|)$; (5.19)

The current at the limit between continuous and discontinuous mode is:

$$I_{0lim} = \frac{V_i T}{2L} D(1-D) = \frac{I_{0lim}}{2|I_0|} D(1-D) \quad (5.20)$$

Therefore, the locus of the limit between continuous and discontinuous modes is given by

$$\frac{1}{2|I_0|} D(1-D) = 1 \quad (5.21)$$

These expressions have been plotted in figure 5.5. The difference in behavior between the continuous and discontinuous modes can be seen clearly.

Principles of operation of the 4-switch topology:

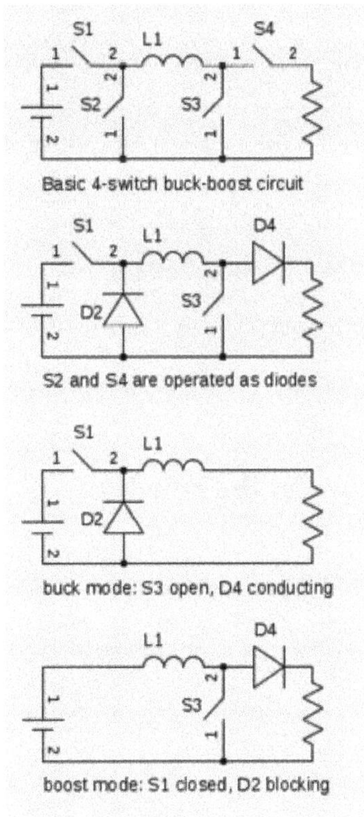

Figure 5.6 4 (Four) Switch Inverter Topology (courtesy client)

The basics of the 4-switch topology. The 4-switch converter combines the buck and boost converters. It can operate in either the buck or the boost mode. In either mode, only one switch controls the duty cycle, another is for commutation and must be operated inversely to the former one, and the remaining two switches are in a fixed position. A 2-switch buck-boost converter can be built with two diodes, but upgrading the diodes to FET transistor switches doesn't cost much extra, while due to lower voltage drop, the efficiency improves.

Non-ideal circuit: Effect of parasitic resistances:

In the analysis above, no dissipative elements (resistors) have been considered. That means that the power is transmitted without losses from the input voltage source to the load.

However, parasitic resistances exist in all circuits, due to the resistivity of the materials they are made from. Therefore, a fraction of the power managed by the converter is dissipated by these parasitic resistances.

For the sake of simplicity, we consider here that the inductor is the only non-ideal component and that it is equivalent to an inductor and a resistor in series. This assumption is acceptable because an inductor is made of one long wound piece of wire, so it is likely to exhibit a non-negligible parasitic resistance (R_L). Furthermore, current flows through the inductor both in the on and off states.

Using the state-space averaging method, we can write:

$$V_i = \bar{V}_L + \bar{V}_s \qquad (5.22)$$

Which are, respectively, the average voltage across the inductor and the switch over the commutation cycle. If we consider that the converter operates in a steady-state, the average current through the inductor is constant. The average voltage across the inductor is:

$$\bar{V}_L = L\frac{d\bar{I}_L}{dt} + R_L\bar{I}_L = R_L\bar{I}_L \qquad (5.23)$$

When the switch is in the on-state V_s=0,

When it is off, the diode is forward biased (we consider the continuous mode operation), therefore $V_s = V_i - V_o$

Therefore, the average voltage across the switch is:

$$\bar{V}_s = D\,x\,0 + (1 - D)(V_i - V_0) = (1 - D)(V_i - V_0) \qquad (5.24)$$

The output current is the opposite of the inductor current during the off-state. The average inductor current is therefore as shown in Fig 5.6:

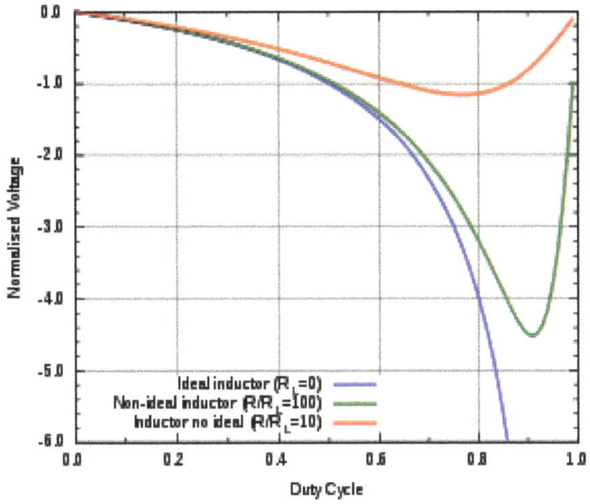

Fig 5.6: Evolution of the output voltage of a buck-boost converter with the duty cycle when the parasitic resistance of the inductor increases.

The average inductor current is given by:

$$\bar{I}_L = \frac{-I_0}{1-D} = \frac{-V_0}{(1-D)R} \tag{5.25}$$

Assuming the output current and voltage have negligible ripple, the load of the converter can be considered purely resistive. If R is the resistance of the load, the above expression becomes:

Using the previous equations, the input voltage ratio with output voltage becomes a comprehensive relation and can be written as:

$$\frac{V_0}{V_i} = \frac{-D}{\frac{R_L}{R(1-D)} + 1 - D} \tag{5.26}$$

If the inductor resistance is zero, the equation above becomes equal to the one of the *ideal* cases. But when R_L increases, the voltage gains of the converter decreases compared to the ideal case. Furthermore, the influence of R_L increases with the duty cycle.

This is summarized in figure 5.6 and illustrated by the changes in the curves of the plots.

<u>Ćuk converter SEPIC converter Split-pi topology</u>

Buck-Boost Converter Operation and Voltage Equation in Discontinuous Conduction Mode (DCM) of Operation.

Here, we need the average value of output voltage <v_o> as V_o in terms of V_i, the input voltage, D, the duty ratio L the inductance, R the resistance, C the capacitance, and T the time period of the switch.

The current plot with respect to duty ratio and total time period is shown in Figure below:

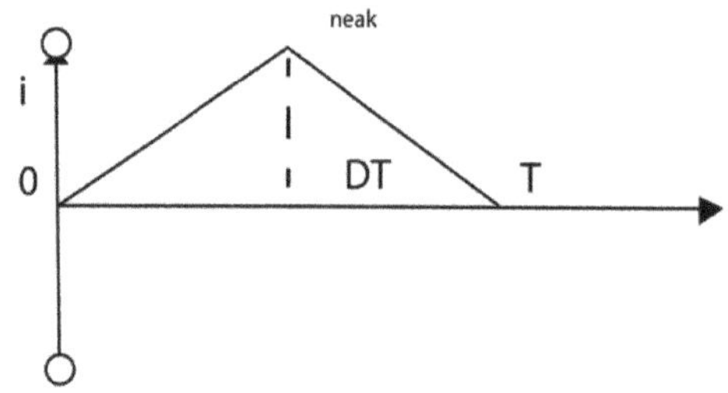

Figure Inductor current plot.

It increases from zero to i_{peak} from time zero to DT and then drops to zero and stays at zero from DT to T.

$$V = L\frac{di}{dt} = L\frac{\Delta i}{\Delta t} \qquad (5.27)$$

When the switch is on and current flows through the inductor from top to bottom

$$V_i = L\frac{i_{peak}}{DT} \qquad (5.28)$$

$$i_{peak} = \frac{V_i DT}{L} \qquad (5.29)$$

Using power balance equation

$$P_{in} = P_{out} \qquad (5.30)$$

$$V_i < i_i > = \frac{{V_o}^2}{R} \qquad (5.31)$$

This entails us to find the value of $< i_i >$ which increases to i_{peak} and drops to zero at time DT and stays there for the entire remaining time period as charge divided by the time:

$$< i_i > = \frac{1/2DT i_{peak}}{T} = \frac{DV_i DT}{2L} = \frac{V_i D^2 T}{2L} \qquad (5.32)$$

Which can be seen to be dependent on the input voltage, duty value, total time period, and inductance value.

Thus, V_o can now be calculated by using this value of $< i_i >$

$$in\ the\ power\ balance\ equationt\ to\ find\ V_o$$

$$V_i \frac{V_i D^2 T}{2L} = \frac{V_o^{\ 2}}{R} \qquad (5.33)$$

Which yields

$$V_o = \sqrt{\frac{R V_i^{\ 2} D^2 T}{2L}} \qquad (5.34)$$

Thus, we get the value of output voltage as:

$$V_o = V_i D \sqrt{\frac{RT}{2L}} \qquad (5.35)$$

Case Study

Control Strategies for Buck/Boost Converters in Electric Vehicle Applications

Esmaeil Gandomkar, Student Member, IEEE, '24

And;

Hemchandra Shertukde, LSMIEEE, '18

Abstract—This paper presents a comprehensive analysis of buck/boost converters, which are pivotal in modern power electronics for efficient voltage regulation in various applications. We explore the operational principles, design considerations, and control strategies of these converters, emphasizing their ability to step down or step up input voltage levels. The study includes a detailed comparison of different topologies, such as the basic buck and boost configurations. Simulation results demonstrate the performance of each topology under varying load conditions and input voltages, providing insights into efficiency and transient response. Furthermore, the paper discusses the implications of emerging technologies, including advanced control techniques and semiconductor materials, on the future development of buck/boost converters. Our findings aim to guide researchers and engineers in optimizing converter design for renewable energy systems, electric vehicles, and other applications requiring reliable power management.

Keywords—DC/DC converter, buck/boost converter, step up and step down, bidirectional power Flow.

Introduction

The growing need for efficient power management solutions has led to the increasing adoption of bi-directional DC-DC converters in applications such as hybrid electric vehicles (HEVs) and electric vehicles (EVs). These converters facilitate flexible power flow, enabling the charging of a low voltage (12V) battery during standard operation in buck mode and assisting high voltage (400V) systems in boost mode during emergencies. A typical configuration includes a half-bridge power stage on the high voltage (HV) side, allowing for seamless energy transfer between different voltage levels.

At the same time, advancements in semiconductor technology, particularly Power devices, have significantly enhanced the performance of power converters. This paper is designed to showcase the high-speed switching capabilities of power MOSFETs.

Utilizing a half-bridge configuration, the converter lets users explore various power conversion topologies, including synchronous and asynchronous buck and boost converters. A low-loss inductor made from "Sendust" material further enhances efficiency by minimizing power losses. Together, these innovations in bidirectional DC-DC converter design and advanced semiconductor technology represent significant strides toward achieving more efficient and flexible power systems, providing valuable tools for engineers and researchers in the field of power electronics.

DC/DC switching converters in HVAC systems are employed to adjust voltage levels for various applications efficiently. When designing these converters, it's essential to consider different levels of integration, ranging from standalone gate driver ICs to fully integrated solutions. Choosing a gate driver design for power devices offers benefits such as short-circuit protection, Dsat protection, under voltage lockout, etc.

In [1], a novel charger design for electric vehicle (EV) batteries is proposed that operates within a wide output voltage range. In [2], the focus is on the integration of a bidirectional converter within a photovoltaic (PV) system to enhance battery management. Various control techniques are also proposed to manage the charging and discharging of the battery effectively. The strategies focus on optimizing energy transfer while maintaining battery health and maximizing system efficiency. In[3], the authors focus on the development of a bidirectional buck-boost converter specifically designed for electric vehicle (EV) applications. Also, this paper presents the design of a bidirectional buck-boost converter, which allows for both step-up and step-down voltage conversion, making it suitable for managing the energy flow between the EV battery and the motor drive.

General Topology

The Half-Bridge converter is a prominent topology in power electronics, especially suitable for applications requiring efficient voltage conversion and non-isolation. This converter consists of two power electronic switches, such as MOSFETs, configured in a half-bridge arrangement. Fig. 5.7 and Fig. 5.8 show the synchronous buck/boost converter, respectively.

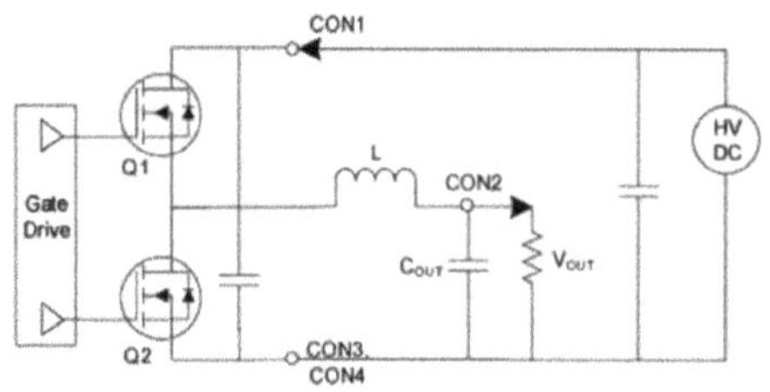

Figure 5.7 Synchronous buck converter.

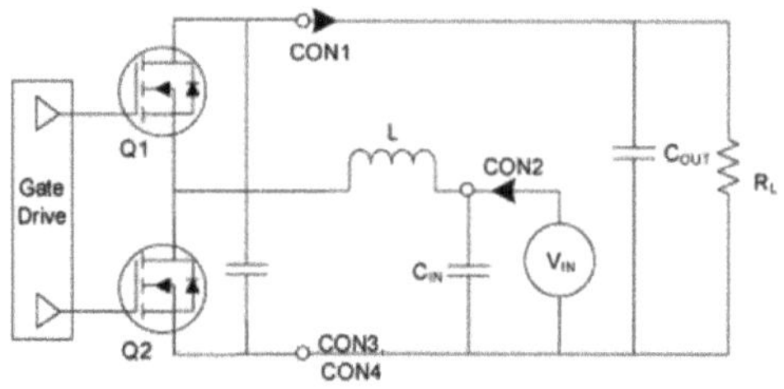

Figure 5.8 Synchronous boost converter.

Mode of Operation (buck/ boost mode)

In the buck mode, the high-side MOSFET (Q1) is controlled using pulse width modulation (PWM) techniques. Complementary PWM determines the overlap between the top and bottom switches. The amount of duty cycle directly influences the converter's performance, allowing for optimized operation under varying load conditions.

In the boost mode of the half-bridge converter, the synchronous switches (Q1 and Q2) are configured to facilitate the conversion of low voltage to high voltage. This mode leverages the buck mode output inductor, which acts as a current source, allowing the converter to operate as a current-fed converter. The full-bridge switches on the high voltage (HV) side can be kept off, utilizing their body diodes for rectification. However, the half-bridge switches are actively employed for step-up during boost mode to enhance efficiency and performance.

Gate driver design [4]

When employing half-bridge configurations, it's essential to generate a high-side bias to control the gate of the high-side MOSFET relative to the switch node. A widely used and cost-effective solution for designers is the bootstrap circuit, which includes a capacitor, a diode, a resistor, and a bypass capacitor. This paper designs the operation of this circuit, the key components involved, and their effects on gate drive performance. It will mainly focus on half-bridge gate drives that utilize drivers without a bootstrap diode, offering designers increased flexibility and lower power dissipation in the gate driver. Fig. 5.9 shows the high side gate driver for the proposed half-bridge configuration.

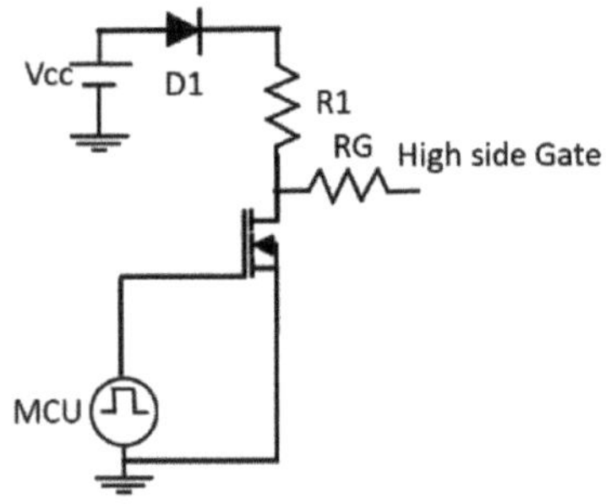

Figure 5.9 High-side gate driver circuit.

As shown in Fig. 5.10, the low-side gate driver can drive the switching power devices in a bidirectional DC/DC buck/boost converter. This high drive strength allows for a fast rise and fall time, which reduces switching losses and ultimately increases the switching converter's efficiency. The analog gate driver also handles voltage on the input pin of the MOSFET for added robustness.

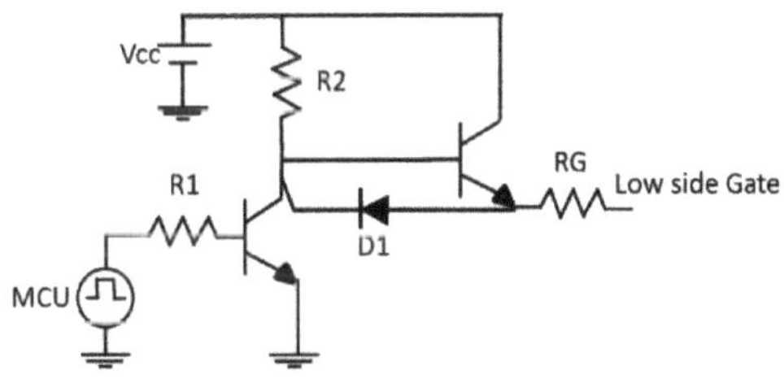

Figure 5.10 Low-side gate driver circuit.

Duty cycle control methodology

The efficiency and performance of the boost mode are fundamentally dictated by the duty cycle of the PWM signals driving switches Q1 and Q2. under a duty-controlled mechanism as shown in figures 5.9 and 5.10. Adjusting the duty cycle allows the amount of energy transferred to the HV side to be precisely regulated, allowing for effective output voltage control. In the buck mode, the output filter, formed by inductor L and capacitor C_{out}, plays a critical role in smoothing the output voltage and current, thereby improving the quality of the power delivered to the load.

PI controller design

The Bi-directional Boost/Buck converter controller utilizes the "C-background/ASM-ISR" framework. In this setup, C code serves as the application's primary program, handling all system management tasks, decision-making, intelligence, and interactions with the host. Assembly code is confined to the Interrupt Service Routine (ISR), where all critical control operations take place, including ADC readings and control calculations.

A single voltage and current feedback loop manage the system. Fig. 5.9 &5.10 illustrate the control blocks; for instance, the voltage controller operates at a frequency of 100 kHz, which matches the PWM switching frequency. The following section details the implemented control strategy.

The sensed output voltage/input voltage/current (Vo/Vin/IL) is compared to the reference voltage/current command (V*/I*) within the voltage/current controller. The output of this controller directly influences the duty cycle of the PWM signals that drive the two half-bridge switches.

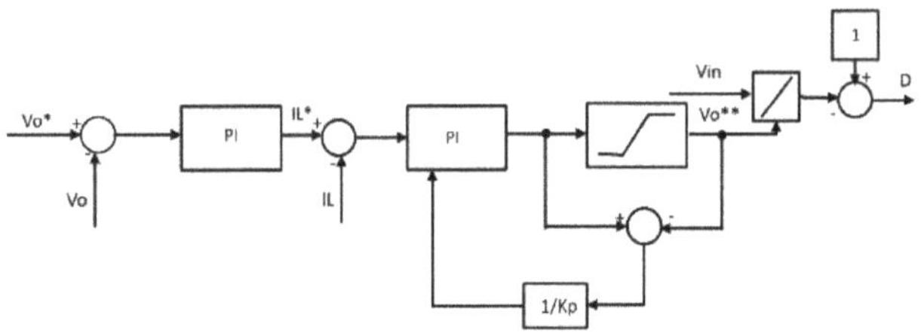

Figure 5.11 PI controller for boost mode

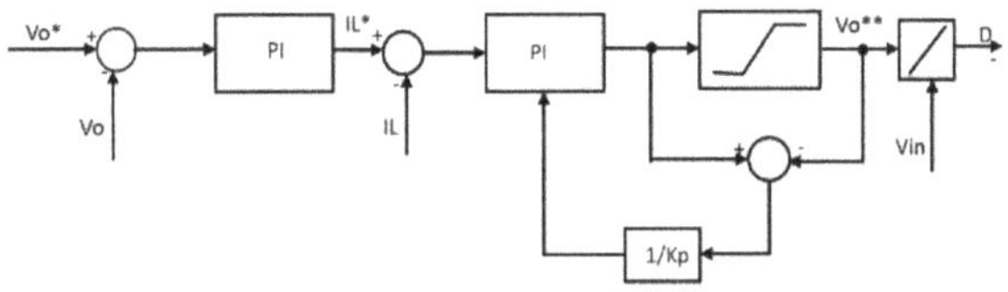

Figure 5.12 PI controller for buck mode

Experimental results

The BOM of the Proposed synchronous buck/boost converter:

2N7002L

50-ohm 0805 resistor

52kohm 0805 resistor

10kohm 0805 resistor

100nF 0805 capacitor

SMMBT2907ALT3G

MMBT2222LT1G

SHBC12-1R0A0028V

7441110 inductors

HHXC500ARA101MJA0G

Step-by-step PCB design with Voltera:

Step 1: Generate a Gerber file of the single-layer PCB in the PCB software (Altium).

Step 2: Upload to the Volterra print PCB software.

Step3: Calibrate the PCB into the Volterra software.

Step 4: print PCB by Volterra on the FR4 plate.

Step 5: bake the printed PCB (in this step, the PCB is ready to use).

Step 6: change the nozzle to the paste nozzle.

Step7: print the top layer by the paste.

Step8: place the component on the PCB (SMD components)

Step 9: bake the PCB up to 180 C to attach the components to the PCB.

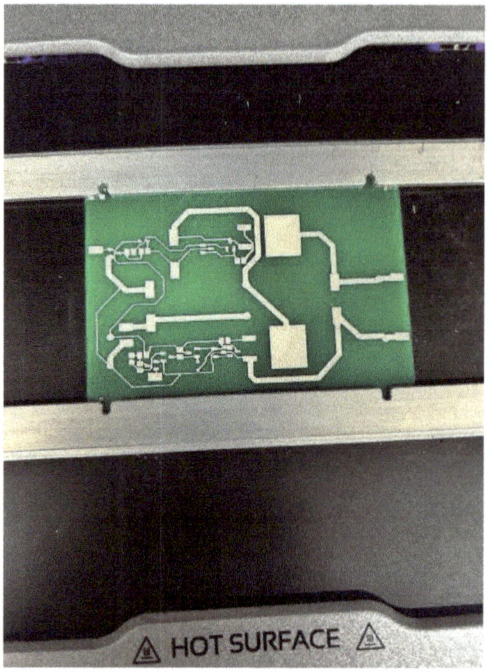

Figurer 5.13 PCB design after printing & baking.

Figure 5.14 Soldering the smc components using Voltera®.

Figure 5.15 The PCB design of buck-boost converter after placing the smc components and Voltera®

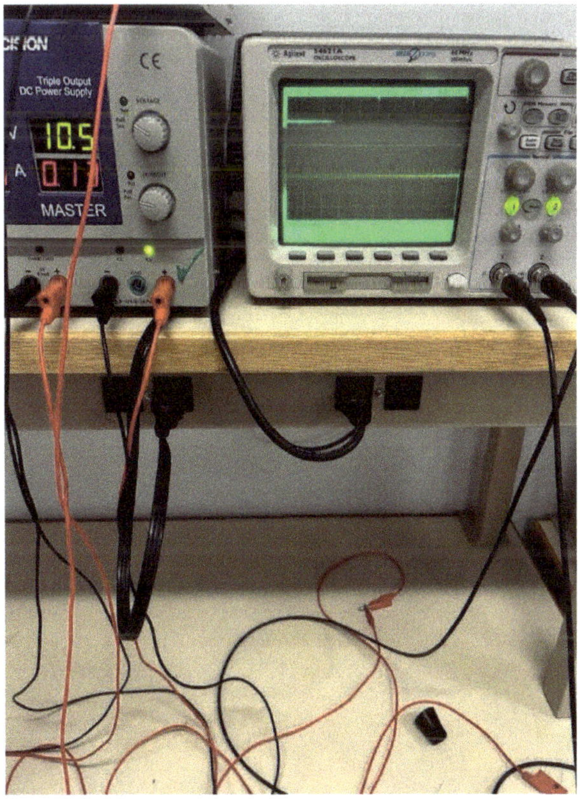

Figure 5.16 Test results with SMC (Boost mode)

Conclusion

In this research, a buck/boost converter has been successfully designed and analyzed for electric vehicle (EV) applications. The study demonstrates the converter's ability to efficiently regulate power flow between the energy storage system and the drivetrain under varying load and voltage conditions. The proposed design achieves high efficiency, compact size, and robust performance, addressing key challenges in EV power management.

The results confirm that the buck/boost converter not only enhances the energy utilization of the battery system but also improves overall system reliability by maintaining stable voltage levels during dynamic operating conditions. This makes it an essential component for modern EV architectures, especially in scenarios involving regenerative braking and multi-voltage energy sources.

Future work

Future work could focus on integrating advanced control algorithms, improving thermal management, and exploring higher power-density designs to optimize the converter for next-generation EV systems further.

Chapter 5 Problems

P.5.1 Design a Buck-Boost DC-DC Converter as suggested in reference below from KIIT, using PSpice or Multisim. Provide output plots and all design parameters clearly in your printed report.

Use the following Reference:

ISSN: 2278-0181 Design of Buck-Boost Converter Using Multisim Software. Moushumi Mishra, Asst. Professor, KIIT University.

Chapter 6

SEPIC Converters

The schematic diagram for a basic SEPIC is shown in Figure 6.1.

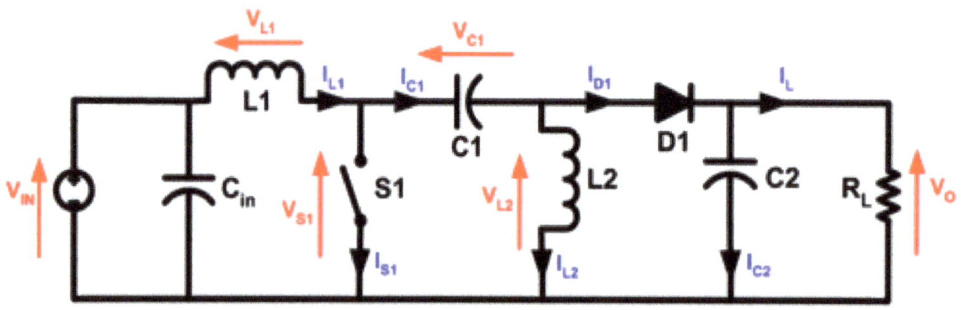

Figure 6.1. SEPIC Converter

As with other switched-mode power supplies (specifically DC-to-DC converters), the SEPIC exchanges energy between the capacitors and inductors in order to convert from one voltage to another. The amount of energy exchanged is controlled by switch S1, which is typically a transistor such as a MOSFET.

MOSFETs offer much higher input impedance and lower voltage drop than bipolar junction transistors (BJTs).

A SEPIC is essentially a boost converter followed by an inverted buck-boost converter, therefore it is similar to a traditional buck-boost converter, but has the advantages of having non-inverted output (the output has the same voltage polarity as the input), using a series capacitor to couple energy from the input to the output (and thus can respond more gracefully to a short-circuit output), and being capable of true shutdown: when the switch S1 is turned off enough, the output (V_0) drops to 0 V, following a fairly hefty transient dump of charge.

SEPICs are useful in applications in which a battery voltage can be above and below that of the regulator's intended output. For example, a single lithium-ion battery typically discharges from 4.2 volts to 3 volts; if other components require 3.3 volts, then the SEPIC would be effective.

Continuous mode:

A SEPIC is said to be in continuous-conduction mode ("continuous mode") if the current through the inductor L1 never falls to zero. During a SEPIC's steady-state operation, the average voltage across capacitor C1 (V_{C1}) is equal to the input voltage (V_{in}). Because capacitor C1 blocks direct current (DC), the average current through it (I_{C1}) is zero, making inductor L2 the only source of DC load current. Therefore, the average current through inductor L2 (I_{L2}) is the same as the average load current and hence independent of the input voltage.

Looking at average voltages, the following can be written:

$$V_{IN} = V_{L1} + V_{C1} + V_{L2} \qquad (6.1)$$

Because the average voltage of V_{C1} is equal to V_{IN}, $V_{L1} = -V_{L2}$. For this reason, the two inductors can be wound on the same core, which begins to resemble a Flyback converter, the most basic of the transformer-isolated SMPS topologies. Since the voltages are the same in magnitude, their effects on the mutual inductance will be zero, assuming the polarity of the windings is correct. Also, since the voltages are the same in magnitude, the ripple currents from the two inductors will be equal in magnitude.

The average currents can be summed as follows (average capacitor currents must be zero):

$$I_{D1} = I_{L!} - I_{L2} \qquad (6.2)$$

When switch S1 is turned on, as shown in Figure 2, current I_{L1} increases, and the current I_{L2} goes more negative. (Mathematically, it decreases due to arrow direction.) The energy to increase the current I_{L1} comes from the input source. Since S1 is a short while closed, and the instantaneous voltage V_{L1} is approximately V_{IN}, the voltage V_{L2} is approximately $-V_{C1}$. Therefore, D1 is opened and the capacitor C1 supplies the energy to increase the magnitude of the current in I_{L2} and thus increase the energy stored in L2. I_L is supplied by C2. The easiest way to visualize this is to consider the bias voltages of the circuit in a DC state, then close S1.

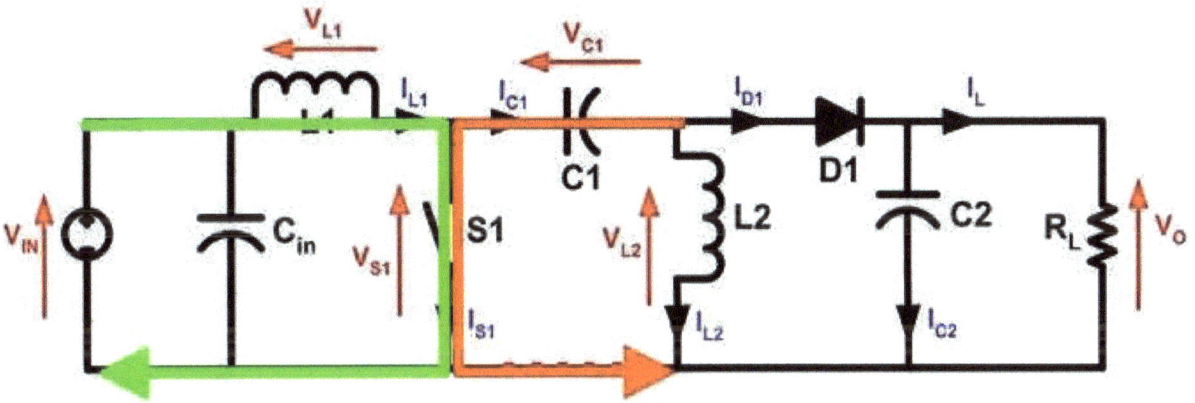

Figure 6.2. Continuous mode of SEPIC converter

SEPIC EQUATIONS AND COMPONENT RATINGS:

Lithium batteries, power-factor converters, and improved low-ESR capacitors are giving a new shine to the classic SEPIC topology. A SEPIC (single-ended primary inductance converter) is distinguished by the fact that its input voltage range can overlap the output voltage. Because SEPIC literature is pretty thin, however, a design engineer not expert in energy converters may feel helpless when asked to design one of these circuits.

This article provides an understanding of basic SEPIC equations, and proposes clear and simple formulas for rating the main components and predicting performance.

Lithium batteries have been very successful, thanks mostly to their impressive energy density. A single lithium cell provides an open voltage of 4.2V when fully charged, and replaces (almost) three of its NiCd or NiMH counterparts. This voltage depends somewhat on residual capacity, and the cell still retains some energy down to 2.7V.

Such input voltage ranges above and below the output of many DC/DC converters, and thereby eliminates the possibility of using an exclusively step-up or step-down type of converter. SEPICs also find application in the power supplies for power-factor converters (PFCs).

Most such circuits use a simple step-up converter as the input stage, implying that the stage output must exceed the peak value of the input waveform. AC inputs of $240V_{RMS}$ ±20%, for example, impose an output of at least 407V, forcing the following converters to work with elevated input voltages. By accepting medium to low input voltages, the SEPIC topology provides a more compact and efficient design. It provides the required output level even if the peak input voltage is higher.

Basic Equations:

The boost (often called step-up) topology (Figure 6.1) is the basis for the SEPIC converter. The boost-converter principle is well understood: first, switch Sw closes during T_{ON}, increasing the magnetic energy stored in inductance L1. Second, the switch opens during T_{OFF}, offering D1 and C_{OUT} as the only path for stored magnetic energy to flow. C_{OUT} filters the current pulse generated by L1 through D1. When V_{OUT} is relatively low, you can improve the efficiency by using a Schottky device with low forward voltage (about 400mV) for D1. V_{OUT} must be higher than V_{IN}. In the opposite case ($V_{IN} > V_{OUT}$) D1 is forward-biased, and nothing prevents current flow from V_{IN} to V_{OUT}.

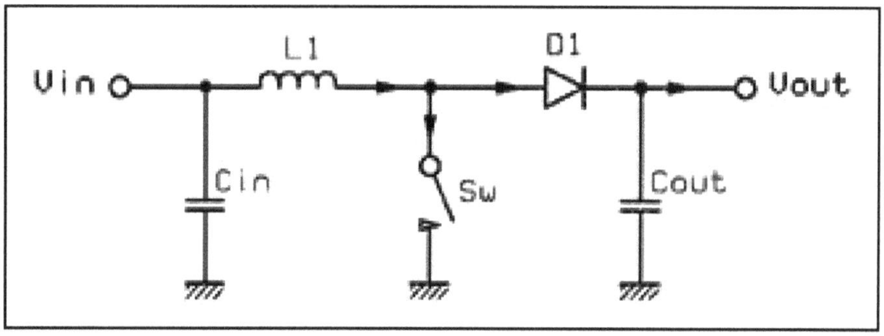

Figure 6.3. This boost-converter topology is the basis for SEPIC power-supply circuits.

The SEPIC scheme of Figure 6.2 removes this limitation by inserting a capacitor (Cp) between L1 and D1. This capacitor obviously blocks any DC component between the input and output. D1's anode, however, must connect to a known potential. This is accomplished by connecting D1 to ground through a second inductor (L2). L2 can be separate from L1 or wound on the same core, depending on the needs of the application. Because the latter configuration is simply a transformer, one might object that a classical fly back topology is more appropriate in that case. The transformer leakage inductance, however, which is no problem in SEPIC schemes, often requires a "snubber" network in fly back schemes. The main parasitic resistances R_{L1}, R_{L2}, R_{SW}, and R_{Cp} are associated with L_1, L_2, S_W, and Cp respectively.

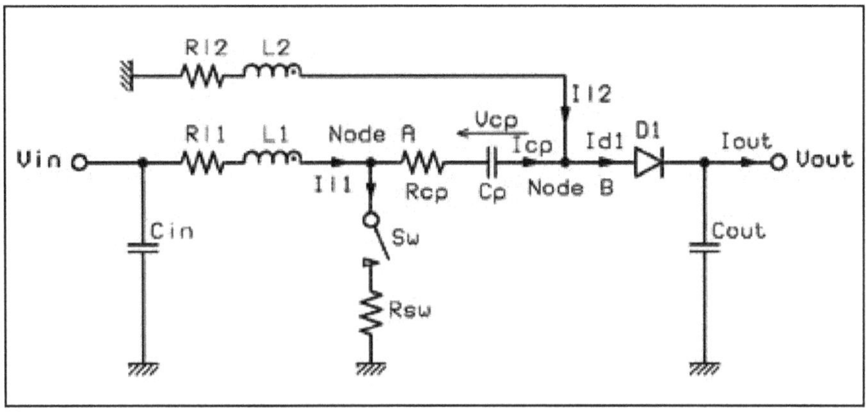

Figure 6.4. An advantage of the SEPIC circuit, besides buck/boost capability, is a capacitor (Cp) that prevents unwanted current flow from V_{IN} to V_{OUT}.

Though it has very few elements, the operation of a SEPIC converter is not so simple to abstract into equations. We assume that the values of current and voltage ripple are small with respect to the DC components. To start, we express the fact that at equilibrium, there is no DC voltage across the two inductances, L1 and L2 (neglecting the voltage drop across their parasitic resistances). Therefore, Cp sees a DC potential of V_{IN} at one side, through L1, and ground on the other side, through L2.

The DC voltage across Cp is:

$$(V_{Cp})_{mean} = V_{IN} \qquad (6.4)$$

"T" is the period of one switching cycle. Call α the portion of T for which switch Sw is closed, and 1-α the remaining part of the period. Because the mean voltage across L1 equals zero during steady-state conditions, the voltage seen by L1 during α T (Ton) is exactly compensated by the voltage seen during (1-α) T (Toff):

$$\alpha\, TV_{IN} = (1-\alpha)\, T\, (V_{OUT} + V_D + V_{Cp} - V_{IN}) = (1-\alpha)\, T\, (V_{OUT} + V_D) \qquad (6.5).$$

V_D is D1's forward voltage drop for a direct current of (IL1 + IL2), and V_{Cp} is equal to V_{IN}:

$$(V_{OUT} + V_D) / V_{IN} = \alpha / (1-\alpha) = Ai \qquad (6.6)$$

Ai is called the amplification factor, where "i" represents the ideal case for which parasitic resistances are null. Neglecting V_D with respect to V_{OUT} (as a first approximation), we see that the ratio of V_{OUT} to V_{IN} can be greater than or less than 1, depending on the value of α (with equality obtained for α = 0.5). This relationship illustrates the peculiarity of SEPIC converters with respect to the classical step-up or step-down topologies. The more accurate expression A_a accounts for parasitic resistances in the circuit:

$$A_a = [V_{out} + V_d + I_{out}\,(A_i\, R_{cp} + R_{L2})] / [V_{in} - A_i\, (R_{L1} + R_{sw})\, I_{out} - R_{sw} I_{out}] \qquad (6.7)$$

This formula lets you compute the minimum, typical, and maximum amplification factors for V_{IN} (A_{amin}, A_{atyp}, and A_{amax}). The formula is recursive ("A" appears in both the result and the expression), but a few iterative calculations lead to the solution asymptotically. The expression neglects transition losses due to the switch Sw and reverse current in D1.

Those losses are usually negligible, especially if Sw is a fast MOSFET and its drain-voltage excursion (V_{IN} + V_{OUT} + V_D) remains under 30V (the apparent limit for today's low-loss MOSFETs).

In some cases, you should also account for losses due to the reverse current of D1, and for core losses due to high-level induction gradients. You can extrapolate the corresponding values of α from Equation (6.2):

$\alpha\ xxx = A_{axxx} / (1+A_{axxx})$, where xxx is min, typ or max. (6.8)

The DC current through Cp is null, so the mean output current can only be supplied by L2:

$I_{OUT} = IL2$ (6.9)

L2's power-dissipation requirement is eased because the mean current into L2 always equals I_{OUT} and does not depend on variations of V_{IN}. To calculate the current into L1 (IL1), express the fact that no DC current can flow through Cp. Thus, the coulomb the charge flowing during α T is perfectly balanced by an opposite coulomb charge during (1-α) T. When the switch is closed (for an interval α T) the node A potential is fixed at 0V.

According to Equation 1, the node B potential is $-V_{IN}$, which reverse-biases D1. Current through Cp is then IL2. When the switch is open during (1-α) T, IL2 flows through D1 while IL1 flows through Cp: $\alpha\ T \times IL2 = (1-\alpha)\ T \times IL1$. Knowing that $IL2 = I_{OUT}$, $IL1 = A_{axxx} \times I_{OUT}$ (6.7)

Because input power equals output power divided by efficiency, IL1 depends strongly on V_{IN}. For a given output power, IL1 increases if V_{IN} decreases. Knowing that IL2 (hence I_{OUT}) flows into Cp during α T, we choose Cp so that its ripple ΔV_{Cp} is a very small fraction of V_{Cp} (γ = 1% to 5%). The worst case occurs when V_{IN} is minimal.

$Cp \geq I_{OUT}\ \alpha_{min}\ T / (\gamma\ V_{INmin})$ (6.10)

The combination of high-frequency controller operation and recent progress in multilayer ceramic capacitors (MLCs) allows the use of small, non-polarized capacitors for Cp. Be sure that Cp is able to sustain the power dissipation Pcp due to its own internal resistance (Rcp):

$Pcp = A_{amin}\ Rcp\ I_{OUT}^2$ (6.11)

Rsw usually consisting of the MOSFET switch drain-to-source resistance in series with a shunt for limiting the maximum current incurs the following loss:

$Psw = A_{amin}\ (1 + A_{amin})\ Rsw\ I_{OUT}^2$ (6.12)

Losses Prl1 and Prl2 due to the internal resistances of L1 and L2 are easily calculated:

$$Prl1 = A_{amin}^2\, Rl1\, I_{OUT}^2 \qquad (6.13)$$

$$Prl2 = RL2\, I_{OUT}^2 \qquad (6.14)$$

When calculating the loss due to D1, take care to evaluate V_D for the sum of IL1 + IL2:

$$PD1 = V_D \times I_{OUT} \qquad (6.15)$$

L1 is chosen so its total current ripple (ΔIL1) is a fraction (β = 20% to 50%) of IL1. The worst case for β occurs when V_{IN} is maximum because DIL1 is maximum when IL1 is minimum. Assuming β = 0.5:

$$L1_{min} = 2\, T\, (1-\alpha_{max})\, V_{INmax} / I_{OUT} \qquad (6.16)$$

Choose a standard value nearest to that calculated for L1, and make sure its saturation current meets the following condition:

$$IL1_{sat} >> IL1 + 0.5\, \Delta IL1 = A_{amin}\, I_{OUT} + 0.5\, T\, \alpha_{min}\, V_{INmin} / L1 \qquad (6.17)$$

The calculation for L2 is similar to that for L1:

$$L2_{min} = 2\, T\, \alpha_{max}\, V_{INmax}/I_{OUT} \qquad (6.18)$$

$$IL2_{sat} >> IL2 + 0.5\, \Delta IL2 = I_{OUT} + 0.5\, T\, \alpha_{max}\, V_{INmax} / L2 \qquad (6.19)$$

If L1 and L2 are wound on the same core, you must choose the larger of the two values.

A single core compels the two windings to have the same number of turns and therefore the same inductance values. Otherwise, voltages across the two windings will differ and Cp will act as a short circuit to the difference. If the winding voltages are identical, they generate equal and cumulative current gradients. Thus, the natural inductance of each winding should equal only half of the value calculated for L1 and L2.

Because no great potential difference exists between the two windings, you can save costs by winding them together in the same operation. If the windings' cross-sections are equivalent, the resistive losses will differ because their currents (IL1 and IL2) differ.

Total loss, however, is lowest when losses are distributed equally between the two windings, so it is useful to set each winding's cross-section according to the current it carries. This is particularly easy to do when the windings consist of split wire for counteracting skin effects. Finally, the core size is chosen to accommodate a saturation current much greater than (IL1 + IL2 + ΔIL1) at the highest core temperature anticipated.

The purpose of the output capacitor (C_{OUT}) is to average the current pulses supplied by D1 during Toff. The current transitions are brutal, so C_{OUT} should be a high-performance Component, like the one used in a flyback topology. Fortunately, today's ceramic capacitors provide low ESR. The minimum value for C_{OUT} is determined by the amount of ripple (ΔV_{OUT}) that can be tolerated:

$$C_{OUT} >= A_{amin} I_{OUT} \alpha_{min} T / \Delta V_{OUT} \quad (6.20)$$

The value of an actual output capacitor may need to be much larger, especially if the load current is composed of high-energy pulses. The input capacitor can be very small, thanks to the filtering properties of the SEPIC topology. Usually, C_{IN} can be ten times smaller than C_{OUT}:

$$C_{IN} = C_{OUT} / 10 \quad (6.21)$$

Overall efficiency η can be predicted from V_{IN} and Aa. The result can be optimistic, however, because it doesn't account for the switch-transition losses or core losses:

$$\eta = V_{OUT} / Aa\, V_{IN} \quad (6.22)$$

Finally, the switch SW and diode D1 should be rated for breakdown voltages, respectively, greater than V_{DS} and V_R:

$$V_{DS} > 1.15\,(V_{OUT} + V_D + V_{IN}) \quad (6.23)$$

$$V_R > 1.15\,(V_{OUT} + V_{IN}) \quad (6.24)$$

As an example, consider component ratings in the following low-power application:

$V_{INmin} = 2.7V$, $V_{INtyp} = 3.5V$, and $V_{INmax} = 5V$, for $V_{OUT} = 3.8V$, $I_{OUT} = 0.38A$, $T = 2\mu S$, and $V_D = 0.4V$. A round of initial estimates gives the following approximate values: L1 and L2 = 47μH, RL1 = RL2 =120mΩ , Rcp = 50mΩ , and Rsw = 170mΩ . Figure 6.3 shows the resulting IL1 and IL2 waveforms at different V_{IN} values.

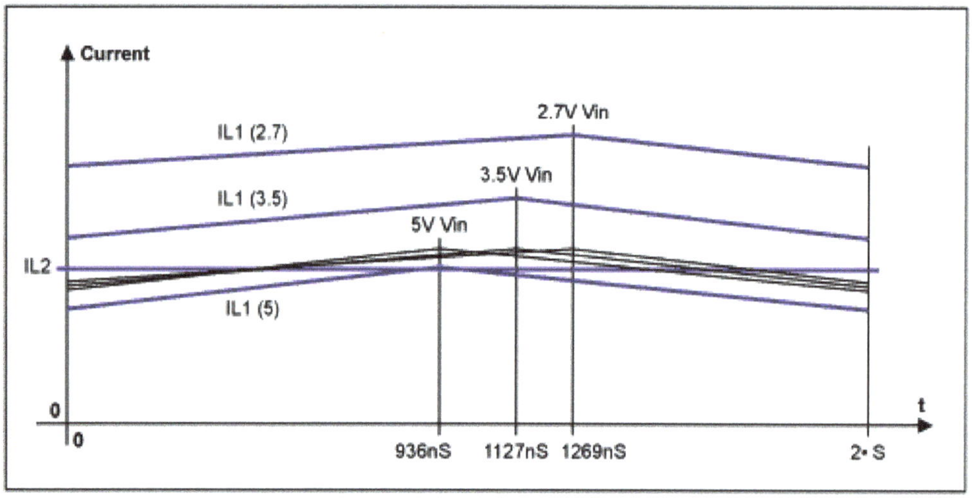

Figure 6.5. In Figure 6.3, the current waveforms through L1 and L2 vary with V_{IN}, as shown.

Using Equation (6.2), you first calculate the ideal amplification factors Ai corresponding to minimum, typical, and maximum V_{IN} as 1.555, 1.2, and 0.84. Using these values in Equation 3, you obtain the more-accurate Aaxxx values of 1.735, 1.292, and 0.88 respectively. The corresponding duty cycles are deduced from Equation 4 as 0.634, 0.563, and 0.468.

The L2 current (IL2) equals 0.38A according to Equation (6.5), and IL1 varies according. To the following:

V_{IN}. Using Equation 6, we obtain IL1 values of 0.659A, 0.491A and 0.334A as V_{IN} varies from minimum to maximum.

We obtain a minimum Cp value of 3.5μF by fixing $\gamma = 5\%$ in Equation (6.7). The voltage rating of Cp is deduced from Equation (6.1). If the input voltage is not to exceed 5V, a 6.8μF ceramic capacitor rated at 6.3V should do the job. Modern MLC capacitors easily meet the expected 50mΩ Rcp, and they easily sustain the 12.5mW power loss deduced from Equation (6.8).

The following parameters are computed at the worst case, which is minimal V_{IN}:

- A 170mΩ switch must dissipate 116.5mW according to Equation (6.9), which allows for the external transistor a SOT23 package or even the smaller SC70.
- Equations (6.10) and (6.11) give losses of 52.2mW and 17.3mW for L1 and L2. We verify here that the copper cross-section of L1 should be larger than that of L2.
- Using Equation (6.12) to calculate the power loss of D1 at 152mW, we see that D1 is the main source of loss. It is, therefore, important to choose an efficient rectifier, if not a synchronous rectifier.
- For L1, Equation (6.13) suggests a minimum value of 28μH, which is close to the estimated value of 47μH. For normal operation with an L1 value of 47μH, Equation (6.14) predicts a peak current of 0.69A. A device rated at 1A provides a reasonable margin. Make sure that D1 can sustain current pulses at high temperature equal to IL1 + I_{OUT} = 1.04A, and a mean current of I_{OUT} = 0.38A.
- Similarly, Equation (6.15) leads to a minimum L2 value of 24.6μH. Again, 47μH is a reasonable value. According to Equation 16, L2 should sustain current peaks of 0.43A.
- For ΔV_{OUT} (V_{OUT} / 100) of 38mV, Equation (6.17) says the output capacitor should be at least 22μF. Equation 18 says 2μF should be sufficient for C_{IN}.
- Despite high-valued parasitic components, Equation 19 predicts a respectable efficiency of 81% for the worst case, in which input voltage is minimum. When transition losses are taken into account, the actual value is a bit lower Voltage

Equation of a Buck-Boost Converter in DCM

Here, we want the average value of the output voltage equal to its RMS value.

This is because we are dealing with DC values, i.e.

$$< v_o > = V_o \qquad (6.25)$$

This Equation is evaluated in terms of V_i, D (duty ratio), L, R, C, and T the period

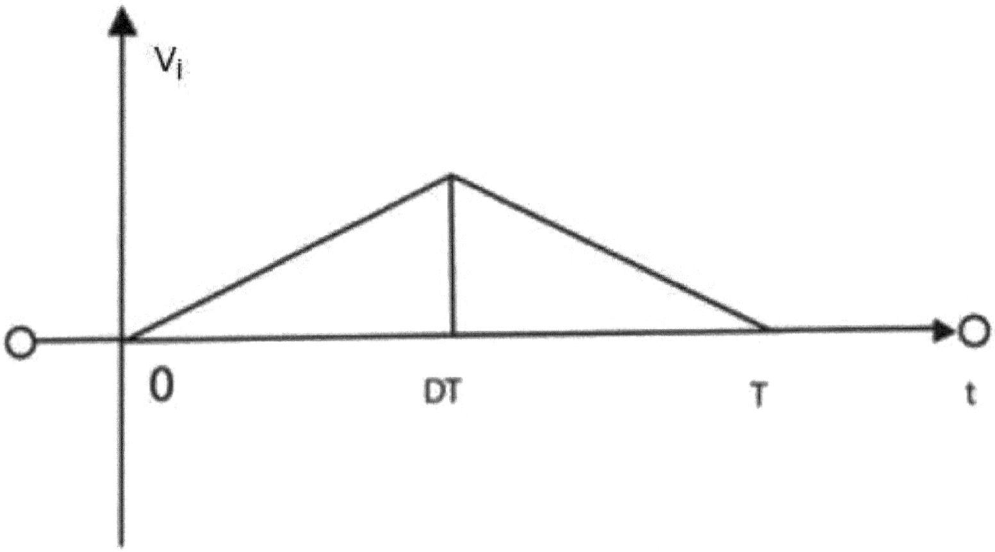

Figure 6.6 Peak Current in DCM

Figure 6.4 shows the peak value of current in DCM. When switch is on the current rises from o to i_{peak}

$$V = L\frac{di}{dt} = L\frac{\Delta i}{\Delta t} \qquad (6.26)$$

In the first segment, when current in the inductor is rising this Equation (6.26) changes to

$$V_i = L\frac{i_{peak}}{Dt} \qquad (6.27)$$

Thus, i_{peak} value by itself is

$$i_{peak} = \frac{V_i Dt}{L} \qquad (6.28)$$

We do not have values of the component L, so we use the power balance equation:

$$P_i = P_o \qquad (6.29)$$

$$V_i < i_i > = \frac{V_o^2}{R} \qquad (6.30)$$

We are trying to solve for the output voltage as required, which yields the average input current during the time from zero to D as rising to the peak value and staying at zero from DT to T

$$< i_i > = \frac{\frac{1}{2}DTi_{peak}}{T} = \frac{DV_iDt}{2L} = \frac{V_iD^2t}{2L} \tag{6.31}$$

This equation, when substituted in the power balance equation yields

$$V_i\frac{V_iD^2t}{2L} = \frac{V_o{}^2}{R} \tag{6.32}$$

Which gives the value for the output voltage of buck-boost dc-dc converter as

$$V_o = V_iD\sqrt{\frac{RT}{2L}} \tag{6.33}$$

We did not end up using the capacitance value as it does not affect the last relationship

Buck-Boost Converter operation and voltage equation in CCM and steady state (average) mode:

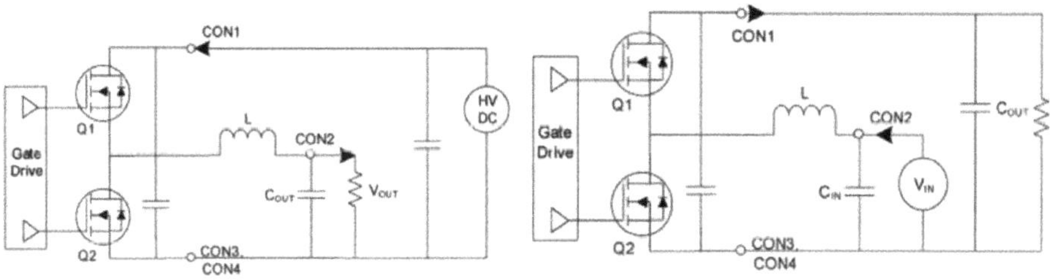

(a) Synchronous buck converter (b) Synchronous boost converter

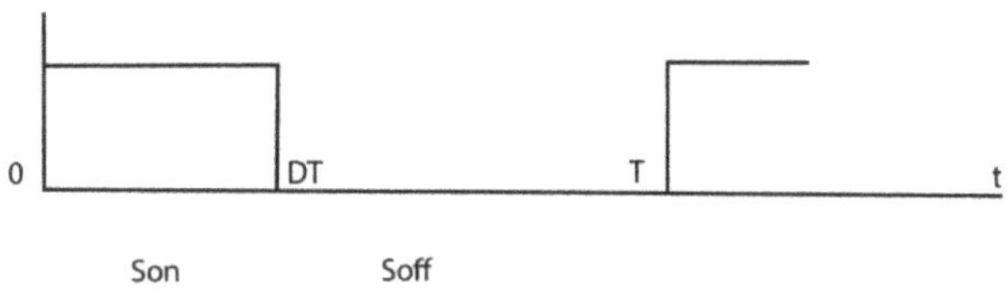

Figure 6.7 Active switch on and off condition

This converter can buck – decrease the voltage and boost or raise the output voltage compared to the input voltage.

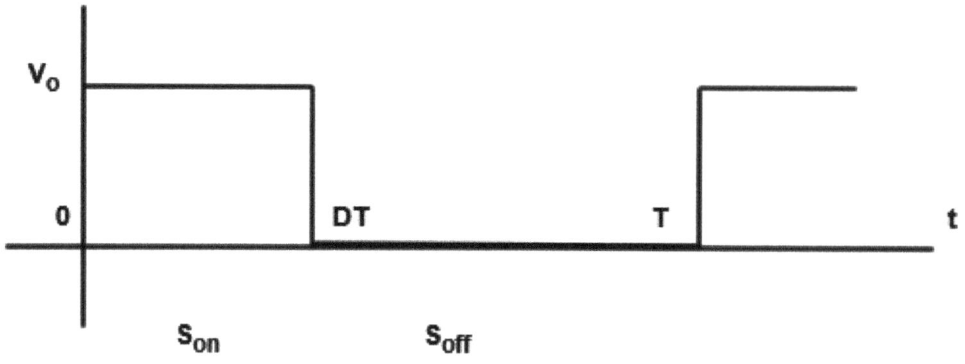

Figure 6.8 Switch time with switch on and off

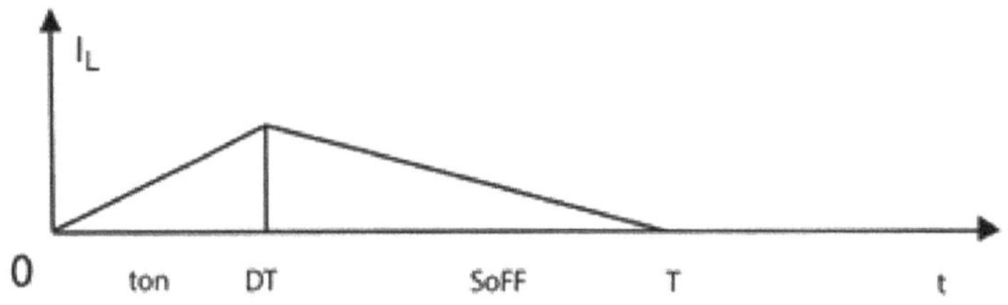

Figure 6.9 of inductor current during switch on and off condition

We assume CCM and $SS_{average}$

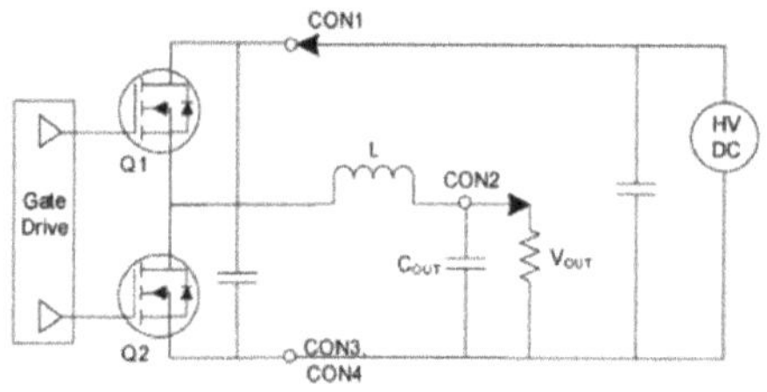

Figure 6.10 Circuit with switch on

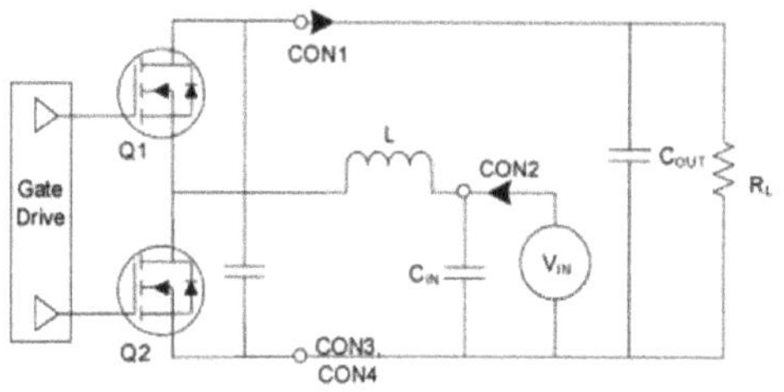

Figure 6.11 Circuit with switch off

Output voltage is in the same direction in both the switch operations, but the polarity at the output is reversed, as shown in the above figures. The current in the inductor is of importance here, which is plotted in the above Figure The slope of the inductor current curve during the switch on condition is

$$\frac{V_i}{L} \tag{6.34}$$

And during the switch off the slope of the inductor current is

$$-\frac{< V_o >}{L} \tag{6.35}$$

The overall change in current over a single time period must be zero. Thus,

$$\Delta i_L = 0 = \frac{V_i}{L}\mathrm{DT} + \frac{-<V_o>}{L}(1-D)T \tag{6.36}$$

$$< v_o > (1-D) = v_i D \tag{6.37}$$

$$v_i =< v_o > \frac{(1-D)}{D} \tag{6.38}$$

$$< v_o >= v_i \frac{D}{1-D} \tag{6.39}$$

This value of D allows us to cause the output voltage to decrease or increase as required with the output voltage polarity reversed from the input voltage.

High Power Boost SEPIC Inverting DC-DC Converters with Fault Protection

This high-power unit generates a 24-volts output from either an 8-volt to 16-volt input. The configuration is shown in the figure above using LT© 3579/3581 DC-DC converters with fault protection. This circuit is capable of creating an output current of 1.1 amps at 24 volts DC. The power switch is integrated into this chip. For LT 3581, it has an output current of 3.3 amps. Because these are monolithic ICs, higher switching frequencies are possible. The switching frequency can be set as free, running between 2.5 KHz to 25 MHz. Or an external clock can be used to synchronize the ports. The above circuit operates at 1 MHz, allowing the use of a physically small inductor. It has an efficiency of 92%.

Applications include boost converters with output voltages of 40 volts. The boost converters are constructed to be protected from output short circuit conditions. These types of DC-DC converters are also constructed for SEPIC DC-DC Converter configuration with low input voltage as low as 4 volts.

The LT 3581 can be used to generate an inverting converter where the input is -12 volts as shown in the figure below. This circuit can generate 625mamps output current. The third circuit below shows and LT 3581 chip to generate a 5-volt output at 0.9 amps from 3 volts to 16 volts input.

All the above circuits also incorporate innovative fault protection features. With the capacitor between the input side to ground and output side to ground can make the fault current a very large value and difficult to control. This large output fault current can be controlled by adding external components.

Such a circuit has a CMOS as shown in figure below.

The PMOS device will turn off thus breaking the path between the input and output path when the fault occurs. The Figure below shows the output current when the fault occurs. The PMOS on and off conditions are for a very short time, after which the device turns on. The same fault current protection circuits can be used to protect against input over voltage conditions and input reverse voltage conditions.

INVERTER-FLYBACK-SEPIC

How to get from Inverter to FLYBACK configuration?

As shown in Figure 1, there is a circuit that shows an input signal connected to a switch followed by an inductor in the first loop. This is connected to the cathode of a diode, followed by a capacitor. At the input, which is a low-impedance source, we have an input capacitor. In this circuit, a positive input voltage gives a negative output voltage. Here, we now add a second inductor coil, as shown in Figure 2, with polarities of low side and high side as shown. Here, also with positive voltage input the circuit yields negative output. We can now modify this circuit as shown in Figure 3. Here, we put the MOSFET to the low side of the transformer circuit. On the output we change the diode connection and polarity of the secondary coil to be connected to the bottom end of the output capacitor. This is called the FLYBACK configuration. We now further modify the circuit in Figure 4, where the end of the low-side coil is connected to the anode of the diode through a capacitor, and the drain end of the MOSFET is connected to the end of the output capacitor. The ratio of the transformer coil in 1:1. This changes the FLYBACK topology to the SEPIC topology. This topology yields the reduction of the ringing at the drain of the MOSFET. The advantage of the FLYBACK circuit in Figure 3 with an additional secondary coil of the transformer to obtain additional voltage at the output, as shown in Figure 5. This circuit also provides isolation in the input power stage.

In addition, the bandwidth is limited by the right half plane zero, and the power is transferred from the input to the output when the switch is open, and the bandwidth is limited by the low pass filter capacitor of the secondary circuit. There is also a possibility of adding a primary auxiliary winding to obtain suitable voltage regulation.

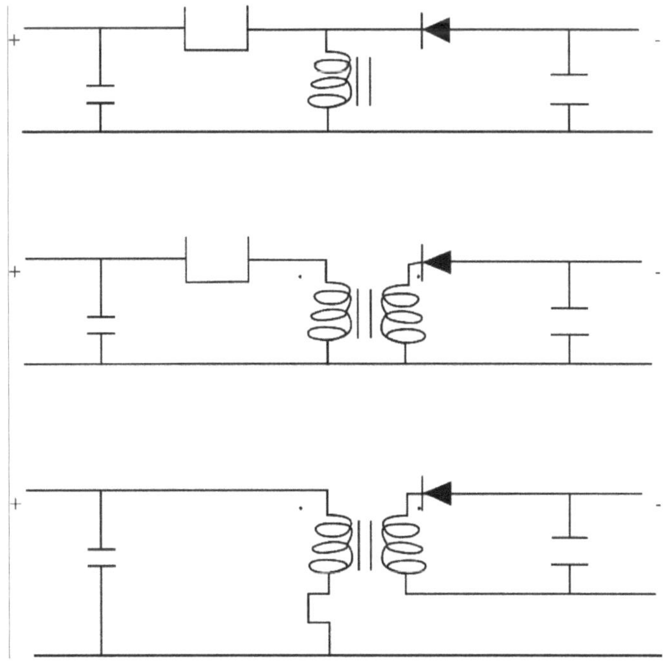

Chapter 7

Filters for HVDC DC-DC Converters

System designers may assume that EMI filters are necessary on the inputs and outputs of DC-DC converters and allow for the cost and space penalty of including them. However, it's worth verifying that the filter is actually necessary as it can degrade electrical performance parameters of the DC-DC, such as efficiency, regulation, and stability. Let's discuss the trade-offs involved.

Output filters in the form of L-C networks are commonly used to reduce the level of switching noise or ripple from a DC-DC converter, see Figure 7.1 showing a typical converter such as in the MEJ series from Murata.

Figure 7.1 Murata MEJ typical dc-dc converter. Courtesy Murata

However, the unfiltered level measured with an oscilloscope probe can be very misleading unless careful attention is paid to the measurement method. The 'scope ground lead can easily pick up extra signal and common mode noise on the output can be seen by the instrument as differential mode. Common mode noise is signal from the outputs with respect to the environment ground, and differential mode is signal across the DC output lines. It is this noise that the following circuitry 'sees' directly and may need the most attenuation. For the filter design, the inductor should be chosen not to saturate at the maximum load current and to have just enough inductance to achieve the attenuation required. Remember that at the 'rated' current of an inductor, inductance may have dropped considerably, so a good margin should be incorporated. Load current steps through the inductance cause voltage transients according to $E = -Ldi/dt$, so unnecessarily large values of L should be avoided.

The DC resistance of the inductor should be low enough so as not to drop excessive voltage at the maximum load current. The mechanical style of the inductor can be important; low-cost, open types such as drum cores can form the primary or secondary of a loosely coupled transformer with other circuitry inducing or picking up noise. If a drum or bobbin style is chosen, the part will have a winding 'start' pin which should be connected to the DC-DC output to provide some electrostatic shielding of this noisy node by the outer layers of winding. The self-resonant frequency of the inductor alone should be high, beyond which its impedance starts to reduce, allowing high-frequency noise to pass through. The capacitor C is often present anyway in the system circuitry but ideally should be placed close to the inductor and DC-DC output and should be a low ESR type. Values are chosen for practicality, but 10 – 100 µH for the inductor and 10 – 100 µF for the capacitor would be typical. The DC-DC data sheet should be consulted for any maximum value of output capacitance allowed. The self-resonant frequency of the combination of L and C will typically be much lower than the DC-DC switching frequency and would not normally present a 'ringing' problem, especially as the circuit is damped by the load resistance. If the DC-DC converter has remote sense connections, beware of connecting them on the load side of a noise filter to try to compensate for the voltage drop across the inductor. The filter would now be within the DC-DC control loop and add phase delay, risking instability. Figure 7.2 shows the output noise performance of a DC-DC from a Murata Power Solutions converter with and without its data sheet recommended filter.

Input filters for DC-DC converters have additional considerations and are often designed to provide common mode as well as differential mode noise attenuation. Differential noise is typically specified in DC-DC data sheets as an RMS or peak-to-peak current through a defined inductance, as the voltage observed depends on the source impedance, which, unlike the load, is unknown to the DC-DC manufacturer. LC filters are again commonly used with the same considerations of inductor saturation, DC resistance, and capacitor ESR. The output impedance and self-resonance of this filter can, however, be critical. If it is incorrectly designed, the filter can interact with the negative input impedance of regulated DC-DC converter, such as the Murata NCM/NCS/NDS series, to produce continuous oscillation. Why is the dynamic impedance of a regulated DC-DC converter negative? For constant load power and converter efficiency, the input power is constant, implying that if input voltage *increases*, for example, the current *decreases* to maintain the constant power, representing the effect of negative resistance. To avoid the oscillation, the input impedance of the DC-DC should be much higher than the output impedance of the filter, which should also be damped so that at resonance, its impedance does not rise close to the DC-DC input value. Damping can be achieved in different ways, but an effective method involves an additional capacitor C_d and resistor R_d in series placed across the input, as shown in Figure 7.3. Detailed component value selection can be found in the literature, starting with *Middlebrook.* Again, excessively large values for L and C can be counterproductive; large inductors can have lower self-resonant frequencies, spoiling attenuation in the MHz region. Large capacitors add to the inrush current perhaps giving problems to the upstream power source. Sometimes, the function of the filter can be more subtle, for example when paralleling DC-DC converter inputs, Murata Power Solutions recommends differential mode filters to prevent noise coupling between the converters, causing potential reliability problems.

Designs often have to control the common mode conducted noise generated by DC-DC converters.

There are standards for the allowed limits of this type of noise, such as EN55022, but technically only relevant to AC-DC converters in information technology equipment. However the standards are often applied to DC-DCs as well. The limits are in the range of hundreds of microvolts across a 50-ohm load in a standard Line Impedance Stabilizer Network (LISN) up to 30 MHz, so filters can be difficult to design - stray impedances and second-order effects in lumped components have to be considered.

There may also be safety considerations; if the input of the DC-DC converter is referenced to line voltage, safety standards limit the allowable leakage current to ground at line frequency, effectively limiting any capacitor value from the DC-DC input to ground and obliging it to be a safety type, typically 'Y' rated at the line voltage. The measured value of mode noise by the LISN is affected by differential and common mode noise. Even with no common mode noise, the standard test set up with an LISN will output half the value of the residual differential mode noise, so filter networks that attenuate both types of noise should be used. Components don't have to be 'lumped', however; common mode chokes can be designed with deliberately loose coupling, effectively adding series mode inductance so that just one inductive component can be used.

Typical types of inductors used for series mode noise suppression are iron powder toroid or drum cores such as the Murata 17/18/19/2200 series, which are low cost and suitable for high current with low inductance. Common mode noise chokes such as the Murata 5000 series are usually high permeability ferrite types in the form of toroid or E-cores. Because in the circuit, the two windings are in antiphase, flux cancels for the normal, circulating operating current so high turns and inductances can be used with little risk of core saturation while giving very high impedance to common mode noise. A comprehensive filter for differential and common mode input noise for a DC-DC converter could take the form of Figure 4, but all elements would rarely be necessary.

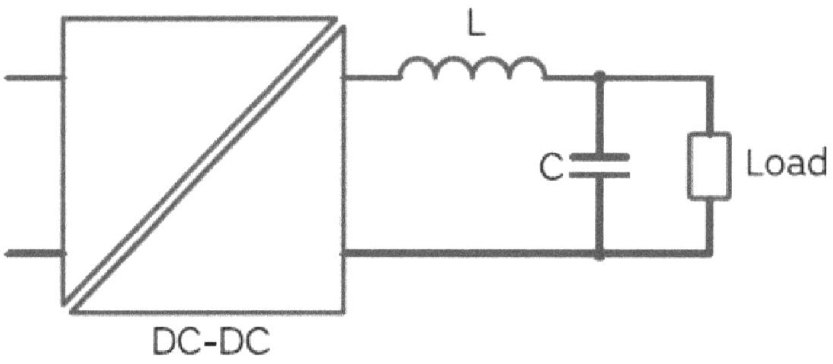

Figure 7.1 (a) Output noise filtering circuit

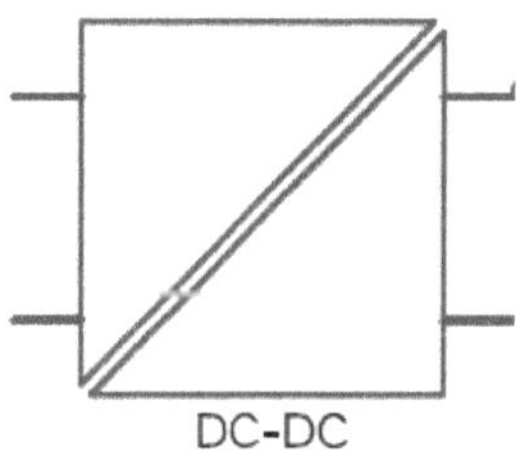

Fig 7.2 without filter

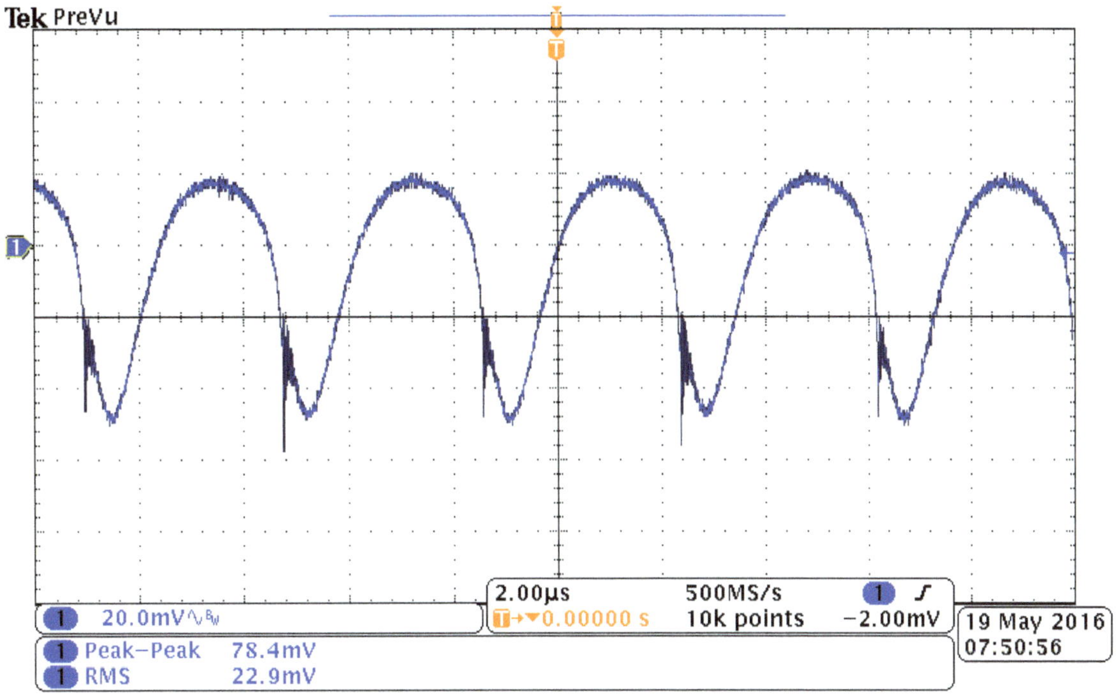

Fig 7.3.a Waveform without filter

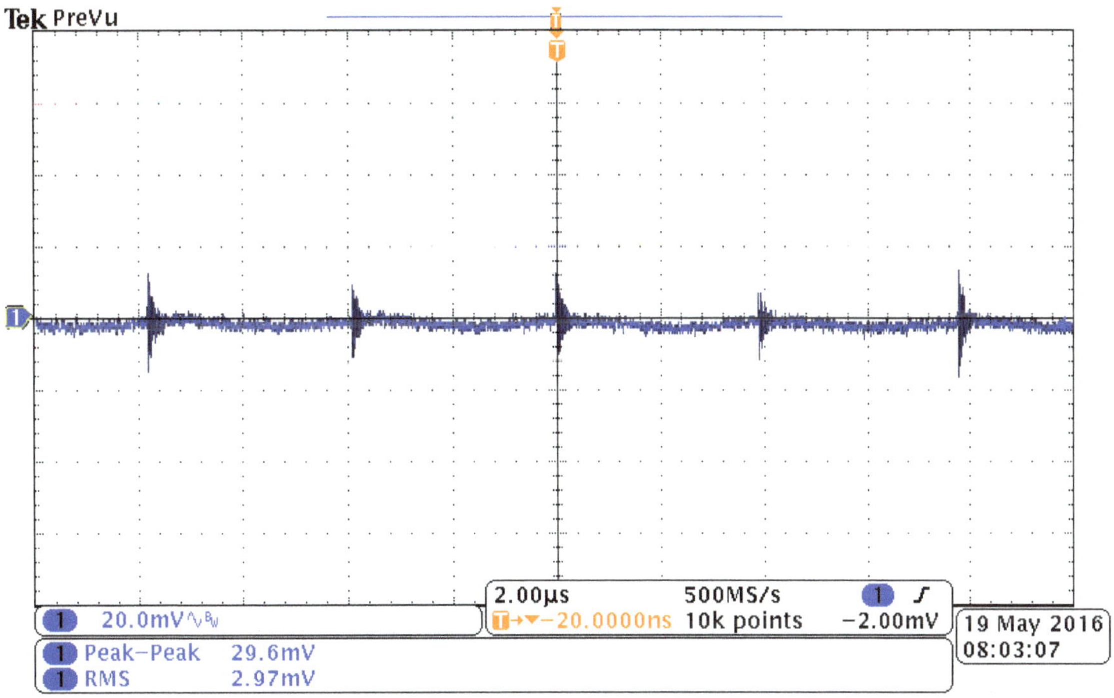

Fig 7.3.b Waveform with filter

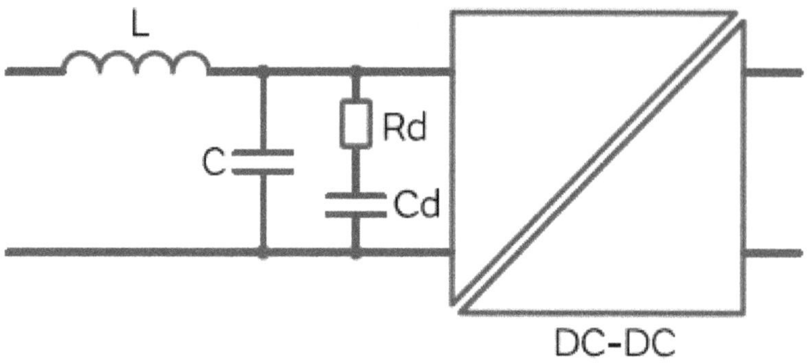

Figure 7.4 Input noise filter to avoid instability

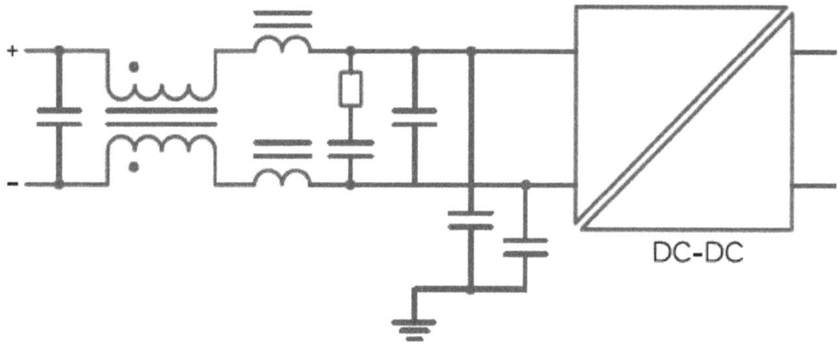

Figure 7.5 Comprehensive DC-DC input noise filter

Chapter 7 Problems

No problems here

Chapter 8

Case Studies

Case Studies as applicable to Solar Energy and other electrical applications SEPIC Controller Using PI controller and MATLAB/Simulink SEPIC with BLDC Motrol Control using fuzzy logic using MATLAB Simulink:

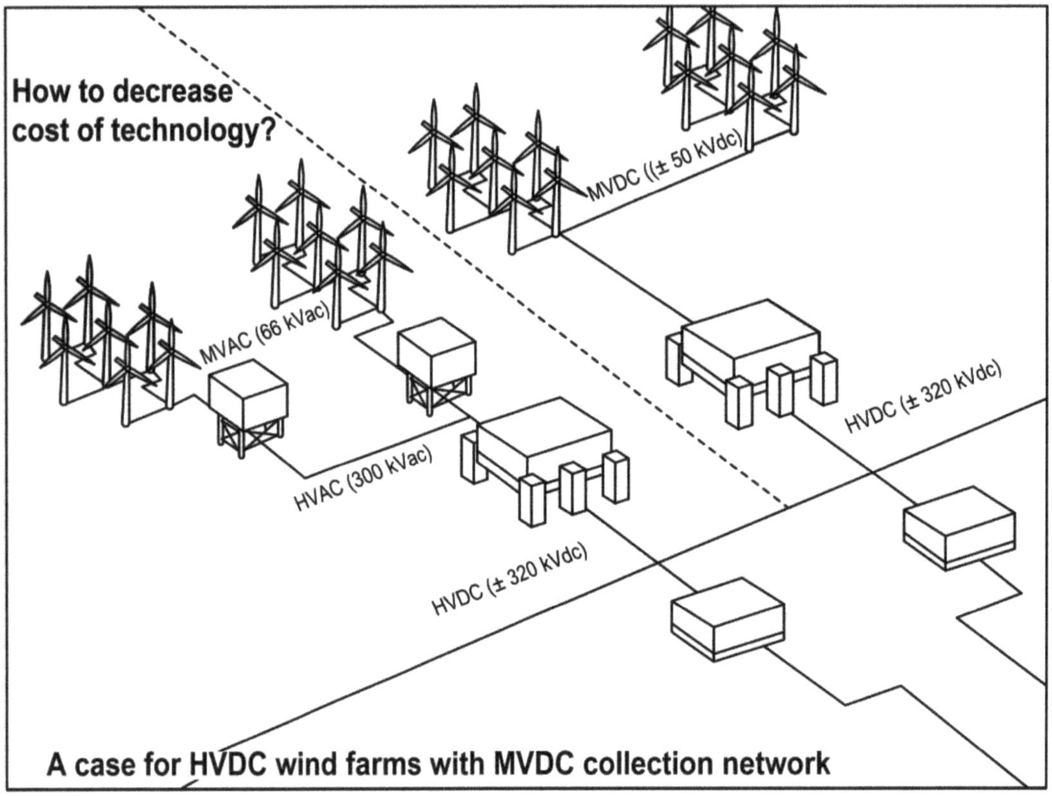

Figure 8.1 High power medium voltage DC-DC converter technology for DC wind turbines

CASE STUDY I – HVDC Wind Farms

Here, we discuss how to reduce cost of technology for HVDC wind farms converter technology and receiving power into MVDC network using off shore wind turbines and DC-DC converters.

The present system consists of HVDC – MVDC Wind turbine/DC-DC converter system on the left and on the right the future solution that can save considerable cost for this new technology.

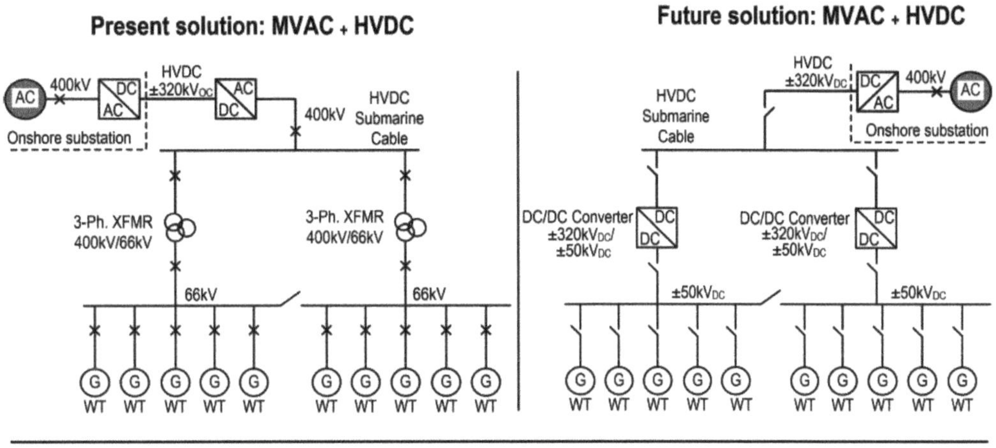

Figure 8.2 Present Solution vs Future Solution

The old system shows standard power conversion system from AC to DC and then DC-DC for a conversion to suitable DC voltage level from HVDC level to MVDC level. The new system shows pure DC wind turbine system at HVDC level with a submarine cable, wind turbines, and suitable network of MVDC.

The annual energy losses are shown below:

The new system with wind farms and DC-DC converter system shows a net saving of 1.2% on losses (cost-wise) and 1.2% savings on bill of material (BOM) cost.

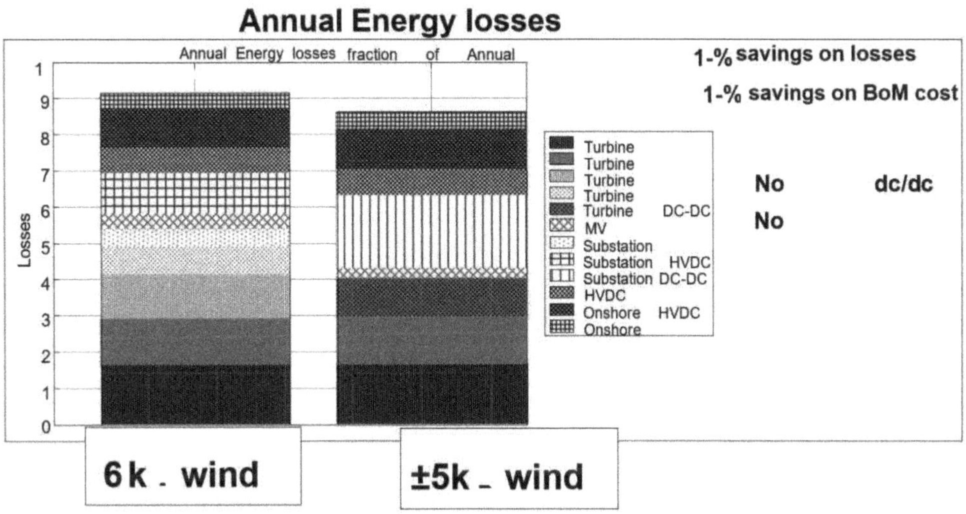

Figure 8.3 Annual Energy Losses for 66 KV AC Wind Farm and +-50 KV DC Wind Farm

The above systems are 66kV AC wind farm vs 50kV DC wind farm. In general, the losses are on a pu basis.

The system does not have a dc-dc converter. There are no standards and there is no experience in this application yet!

The requirements for the system is shown below in Figure 8.4:

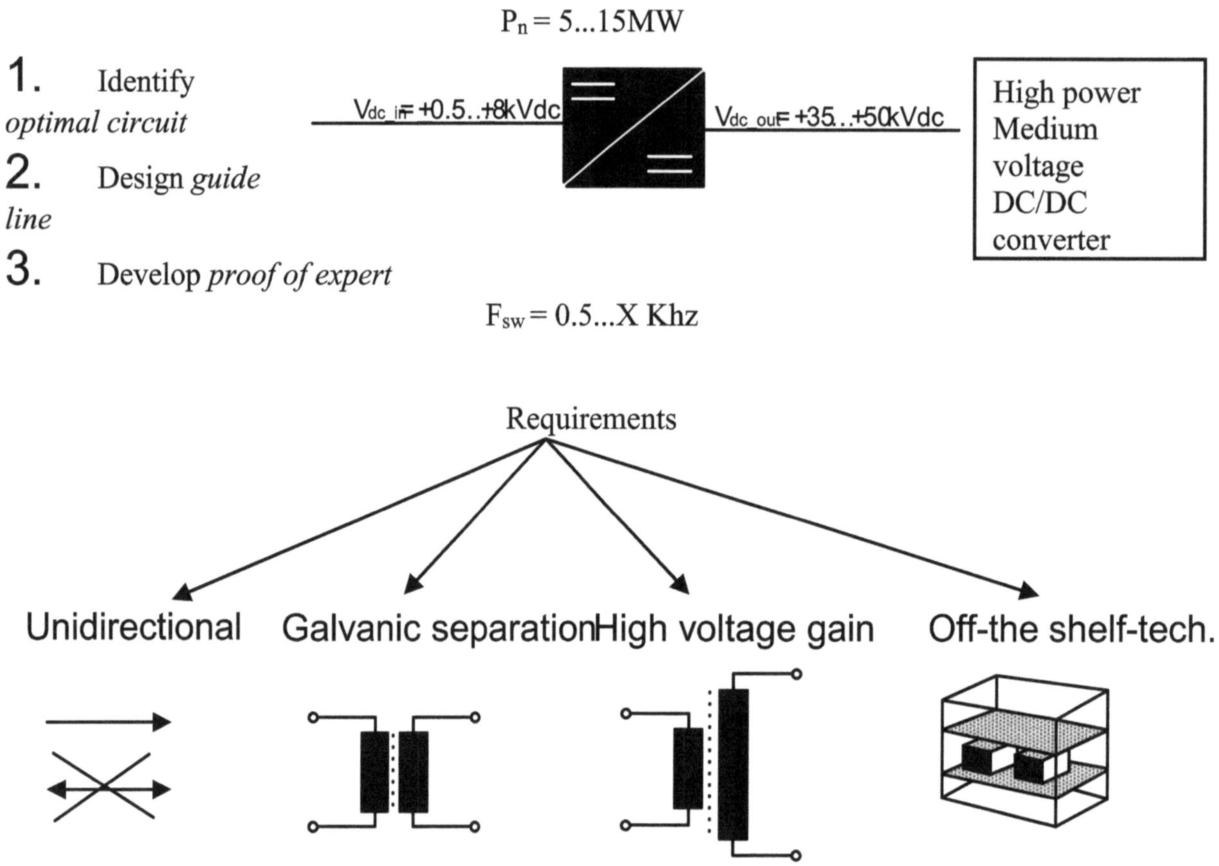

Figure 8.4 Steps of Main Objective: Identify, Design, and Develop.

Some very salient requirements were:

1) The system is unidirectional
2) Need galvanic separation
3) High voltage gain
4) Off-the-shelf systems to lower some costs

Thus, we have the latest system, as shown below in Figure 8.5:

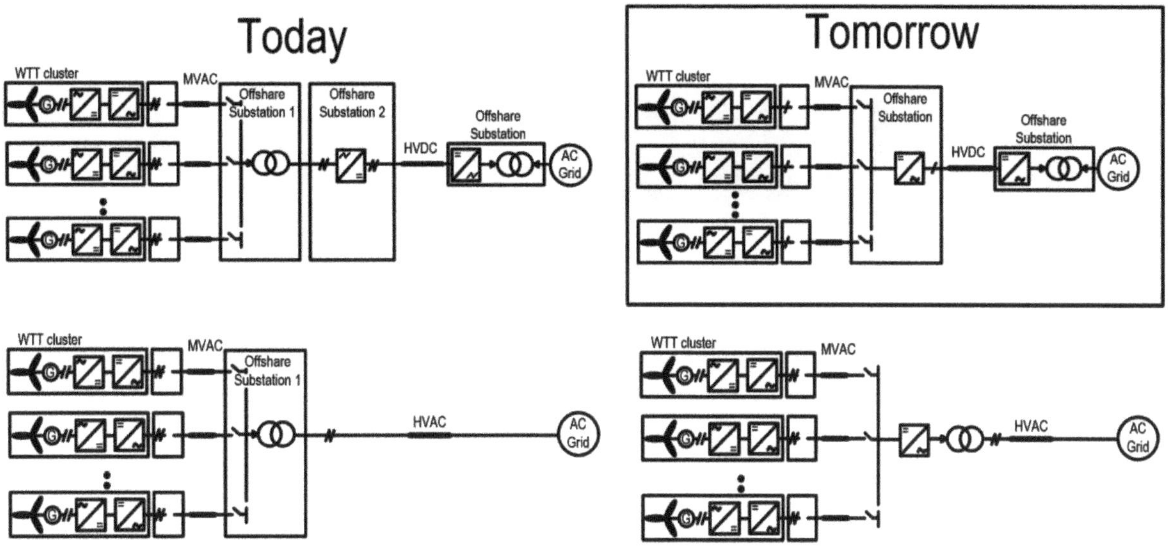

Figure 8.5 Selected Wind Farm Configuration.

In detail, this looks as follows as in Figure 8.6:

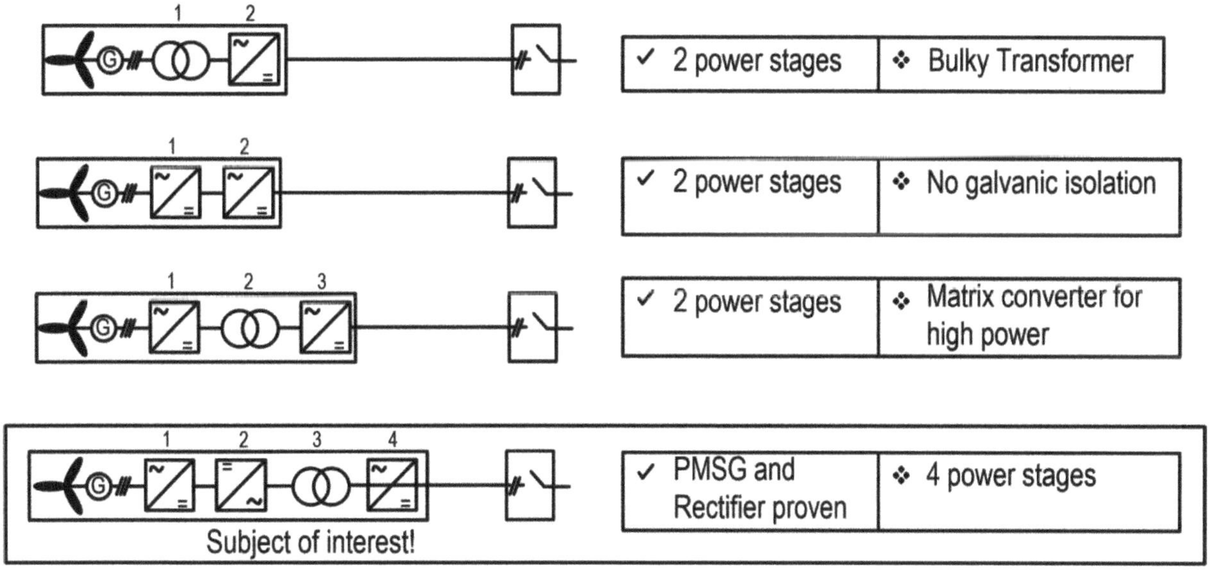

Highest priorities: Circuit design, semiconductors Lower priority: Transformer design

Figure 8.6 In real life Turbine Converter Configuration (courtesy client)

The different configurations change depending on the size and KW level of the system. The most economical with intentions to provide the highest priority to circuit design and low priority for transformer design, which has been in place for decades, indicates the last system is PMSG proven and rectifier driven with low loss and high gain. Such systems are of interest as well to obtain maximum efficiency for the system that is suitable for wind turbines. It shows the wind turbine followed by two-stage converters sandwiched between them is non-galvanic connection effected with the use of a suitable transformer.

The selection of different topologies is shown below in Figure 8.7:

There are thirty-seven different topologies like

- Full Bridge
- Matrix
- Cascaded Bridge
- Dual Active Bridge
- MMC
- Single Active Bridge (SAB)
- Series Resonant Converter

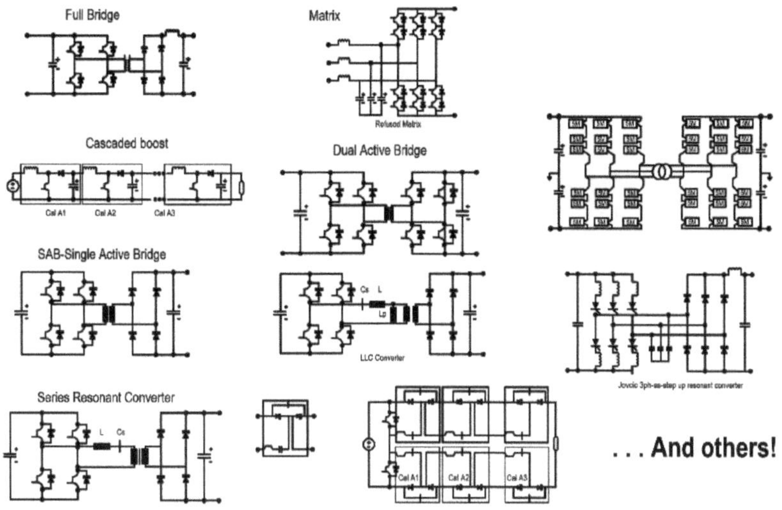

Figure 8.7 Selection Process from 37 Topologies of MMC

These topologies can be arranged together in switching point of view and configuration point of view. Some configurations, as provided by some demonstrators, as shown below in Figure 8.8:

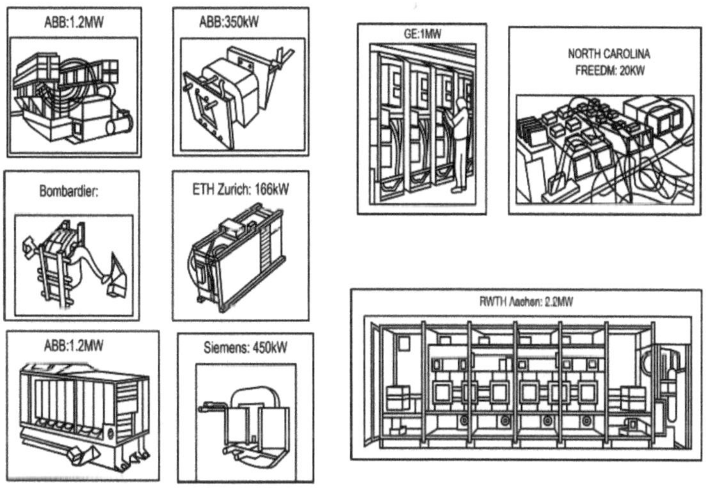

Figure 8.8 Survey of Demonstrators

Some are for Traction applications, as shown above, and others are for solid-state transformer configurations.

We will discuss design drivers for DC Wind Turbines as shown below in Figure 8.9:

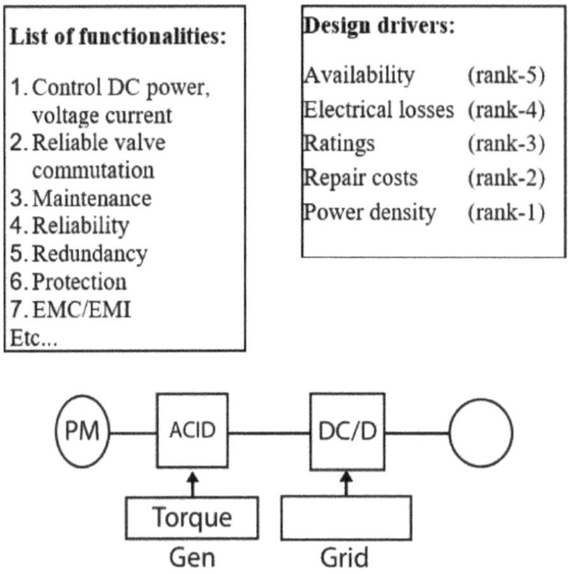

Figure 8.9 Design Drivers for DC wind turbines.

The different and salient functionalities that are necessary cover a large number like:

Control of DC power and voltage and current as well to ensure proper delivery of energy as specified by the system. Reliable switch commutation that needs sufficient and regular maintenance. This often leads to the reliability of the Wind Turbine system, removing redundancy, increasing protection, and other serious failures associated with EMC/EMI. In achieving these parameters, the design drivers get guided by the availability of the different assets, resulting in losses that need to be kept at a minimum to increase efficiency, proper ratings to suit the application of wind turbines, the eventual repair and maintenance cost and the resulting appropriate power density in the region under consideration for which the energy is harnessed.

The DC-DC converter link is of importance in this discussion as it relates to grid converter control moving from HVDC to MVDC network. The AC-DC conversion is monitored from the prime mover side or generator to provide torque control and generator control. To smoothen this changeover, a modularity control process is adopted, as shown below in Figure 8.10:

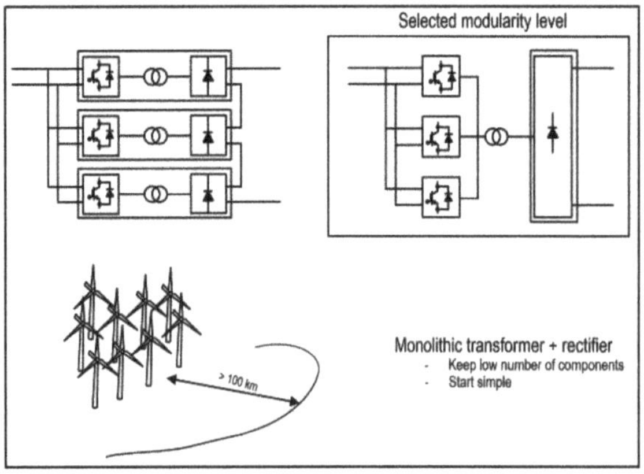

Figure 8.10 Different Modularity Levels

The selected modularity level at the converter level is preferred and is highlighted with a bolded background above. This keeps the cost low as a single transformer will be encountered at a given distance may be several 100 kilometers from the shore, so we will keep one monolithic transformer with high frequency in KHz in the network and a simple rectifier to keep the cost down and resulting failures simple and solvable without much interdependency.

No longer bulky 50z Hz transformer! Savings in volume and weight!
Volume

$$A_{core}A_{wdg} = \frac{\sqrt{2}}{\pi} \frac{P_t}{K_w B_{max} J_{rms} f_{sw}}$$

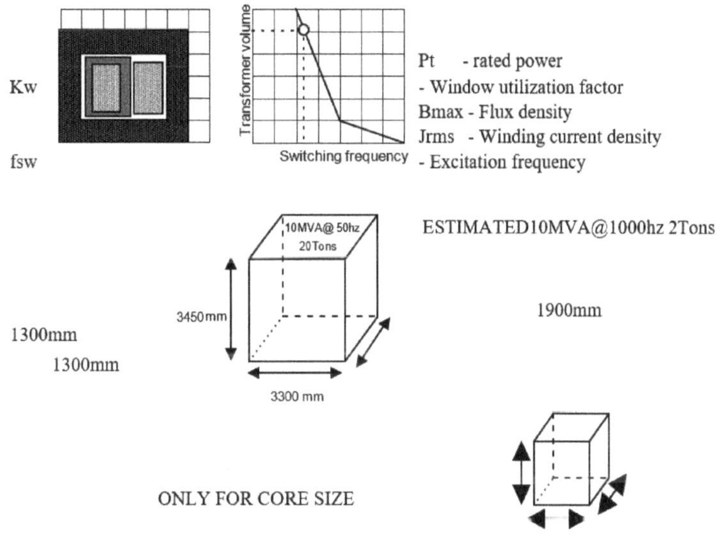

Figure 8.11 Size change due to high-frequency transformer

So, we no longer have a bulky 60 Hz transformer, and the related cost of the transformer core (size) is reduced considerably. An increase in frequency to 1Khz can bring the size of the transformer by 8-10 times resulting in considerable savings. So, a proposed system solution is shown below in Figure 8.12:

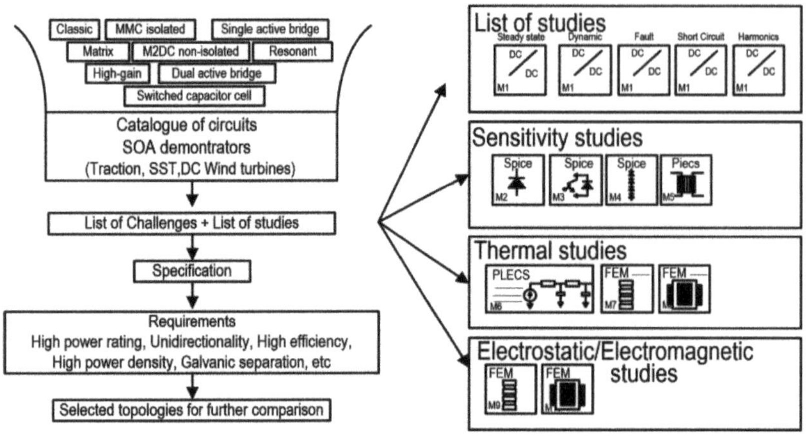

Figure 8.12 Proposed Methodology of the Future System

The above Figure 8.12 categorically describes the process in terms of survey of circuits, provision of demonstrators, choice of topology, and once this is set into a working system to conduct different tests, studies, steady state operation, dynamics and efficiency of the system, response to faults, etc. This followed by sensitivity studies on diodes, IGBTs/IGCTs etc., followed by systematic EMI and ESI studies with a proper and suitable methodology to prove the efficacy of the proposed system.

The turbine converter circuit selection is done by assessing the earlier studies, as shown below in Figure 8.13:

Simulation study results: overview of component loads

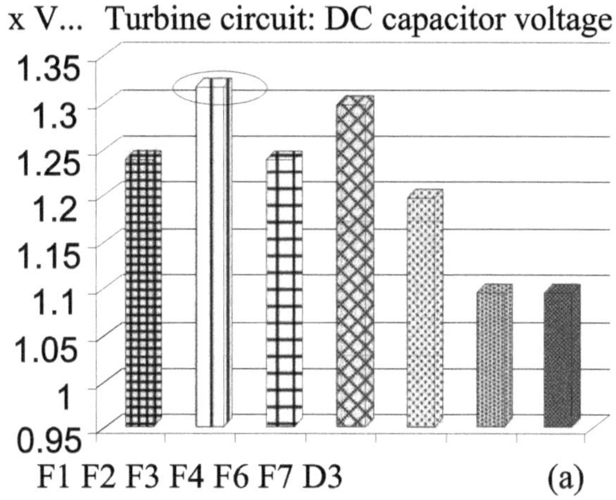

(a)

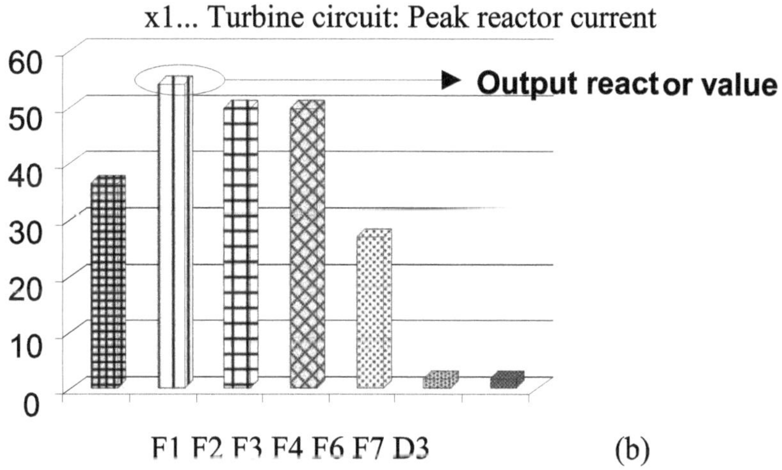

(b)

Studies for Characterization of Electrical of DC Collection System in Offshore Wind Farms Turbine output current and short circuit fault current

(c)

Turbine and offshore substation converter 1 MVDC pole voltage

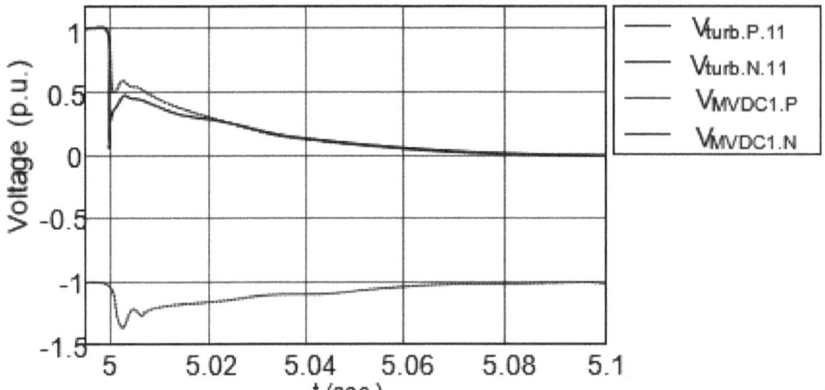

(d)

Figure 8.13 Turbine performance for different component loads as in parts (a-d).

This also completely depends on the collector network and uses causes like short circuit, surges, and transient response as shown above. This will help in the proper design of the converter for a baseline design to assess if the topology selected is appropriate. Figure 8.14 below illustrates a full bridge converter and followed by a monolithic transformer.

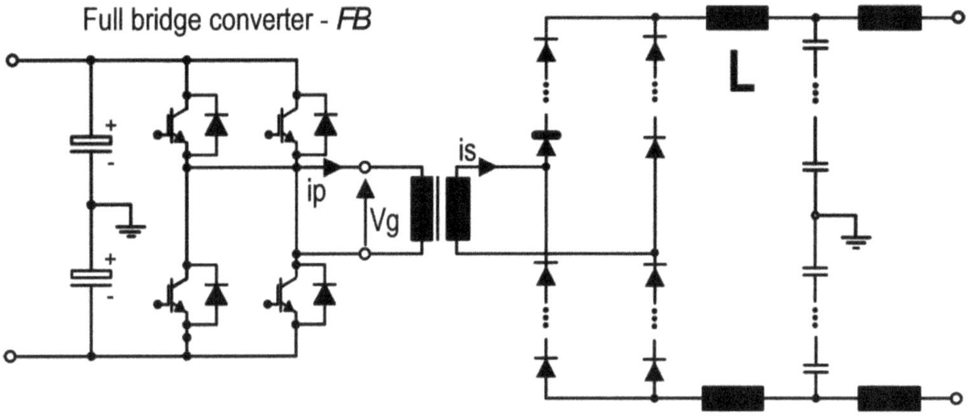

a) FB

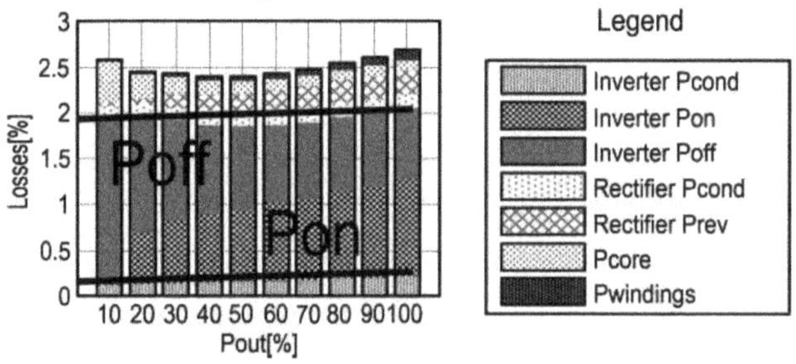

Figure 8.14 Hard switched Full bridge converter

And a rectifier with the series connected diodes for a DC output. If we conduct loss analysis on this configuration it will be less than 3% of the rated. The next choice is a single active converter as below in Figure 8.15:

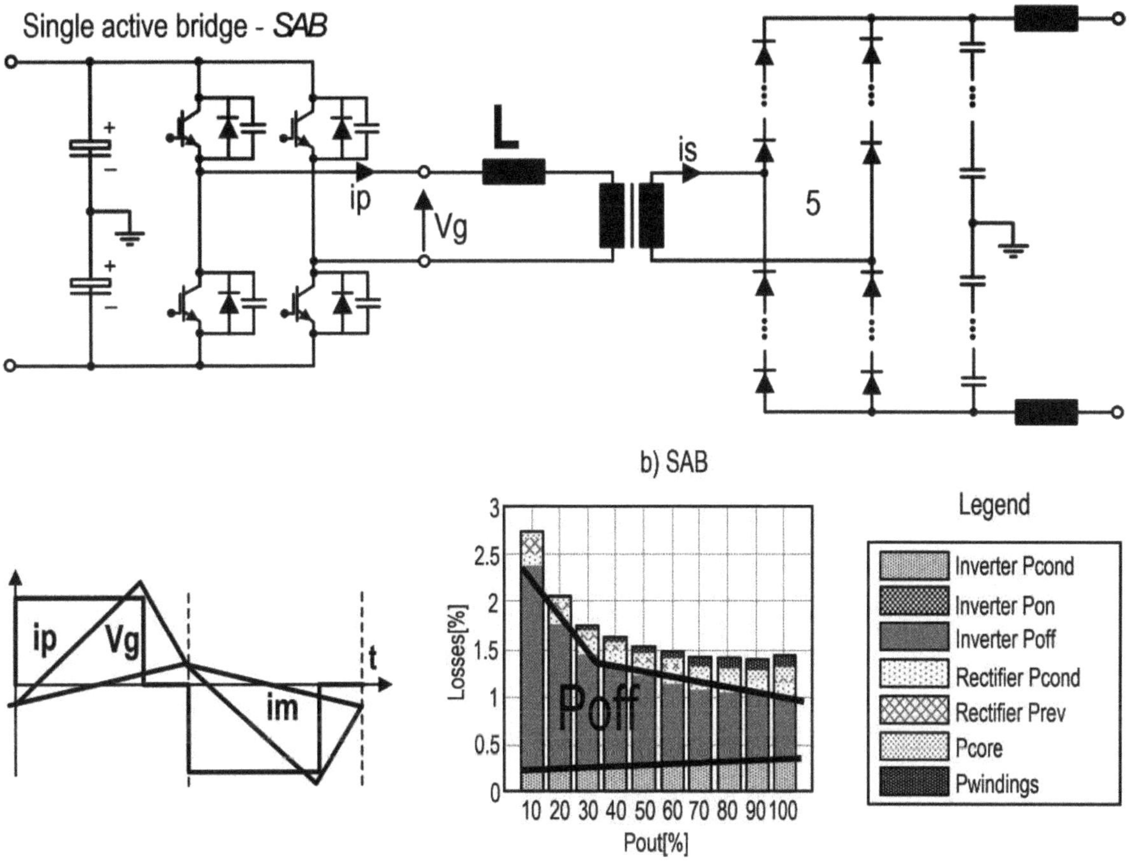

Figure 8.15 Single Active Bridge

In this configuration, as in Figure 8.15 we can improve on the efficiency as the losses are further reduced.

If you also observe the increase in voltage levels at different fault levels and conditions, it is important in respect of the design of the converter as well.

Failure mode effect analysis Concept selection with Pugh matrix:

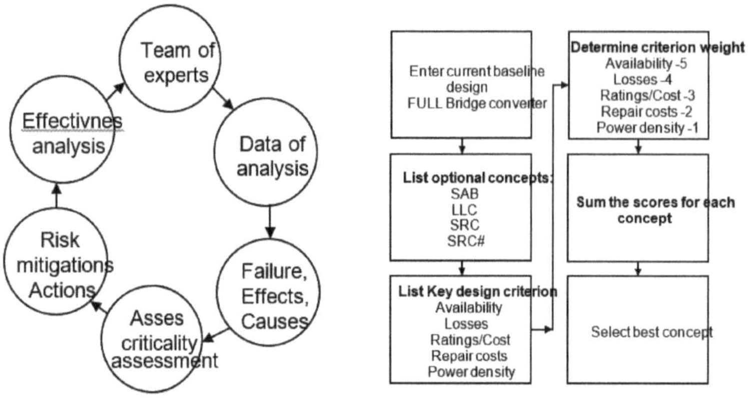

Figure 8.16 Proposed methodology for the future system

The Pugh Matrix methodology is used for selection of a concept topology. This is considerably helpful in conducting failure analysis to help in selection of appropriate topology. From 37 different topologies, five were selected as shown below in Figure 8.17. The first one was for a single-phase system, unidirectional, galvanic separation with a monolithic transformer.

- From 37 topologies, down-select to 5 topologies
- Characteristics: single phase, unidirectional, galvanic separation, monolithic transformer

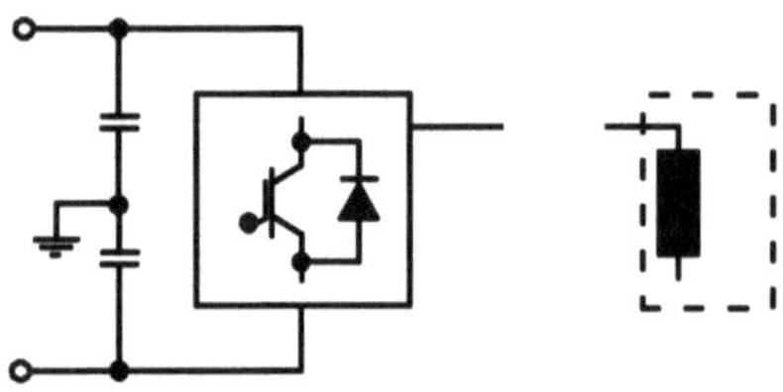

Figure 8.17 Selection of Down Circuits

The next one was a hard-switched full bridge converter, as shown below: It has a full bridge inverter, a monolithic transformer, and a series of connected diodes to create the possible topology. The losses in this case, as indicated earlier, are less than 3% as expected.

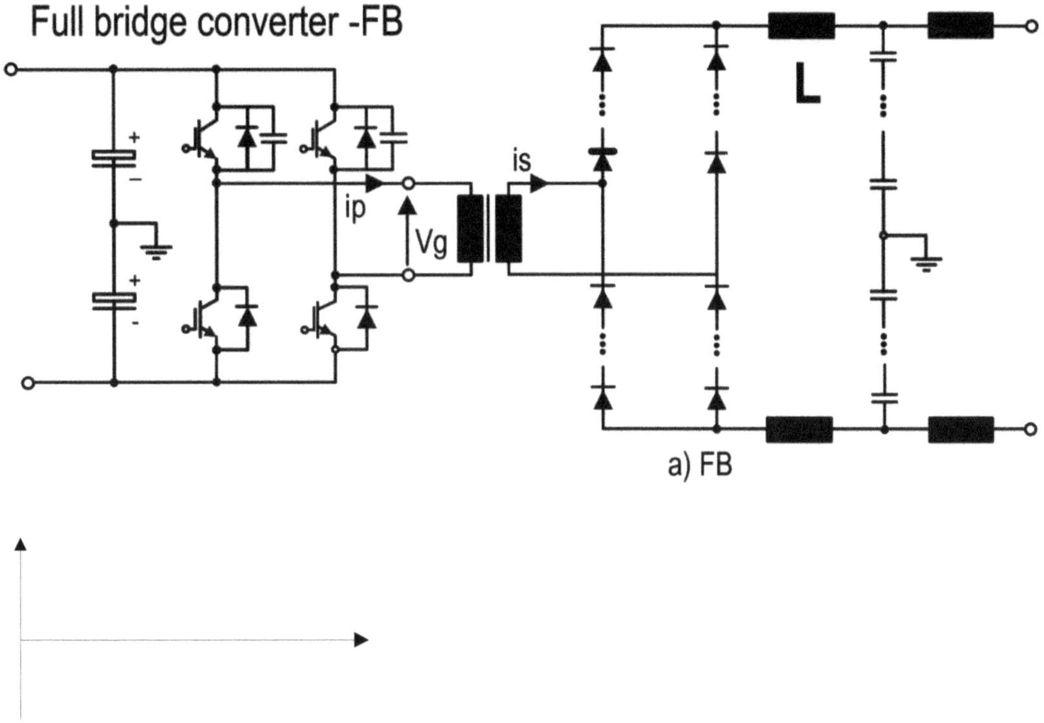

Figure 8.18 Hard-Switched Bridge Converter Circuit Performance

The next option would be a single active bridge topology, as shown below in Figure 8.19:

Figure 8.19 Single Active Bridge Topology and performance at different loads. The current waveform here as expected from the previous trapezoidal to triangular, thus reducing or eliminating the turn-on losses, thereby increasing the efficiency. But for large HVDC systems, such a design would need an exceptionally low leakage inductance for the transformer, which is an impractical situation. Thus, there are some advantages in improving efficiency, but such designs may not be practical.

In addition, besides the practical limitations on high-frequency transformers, the other aspects are the non-ideal situations set in by the stray losses due to high-frequency signals. Such modalities are shown below in Figure 8.20:

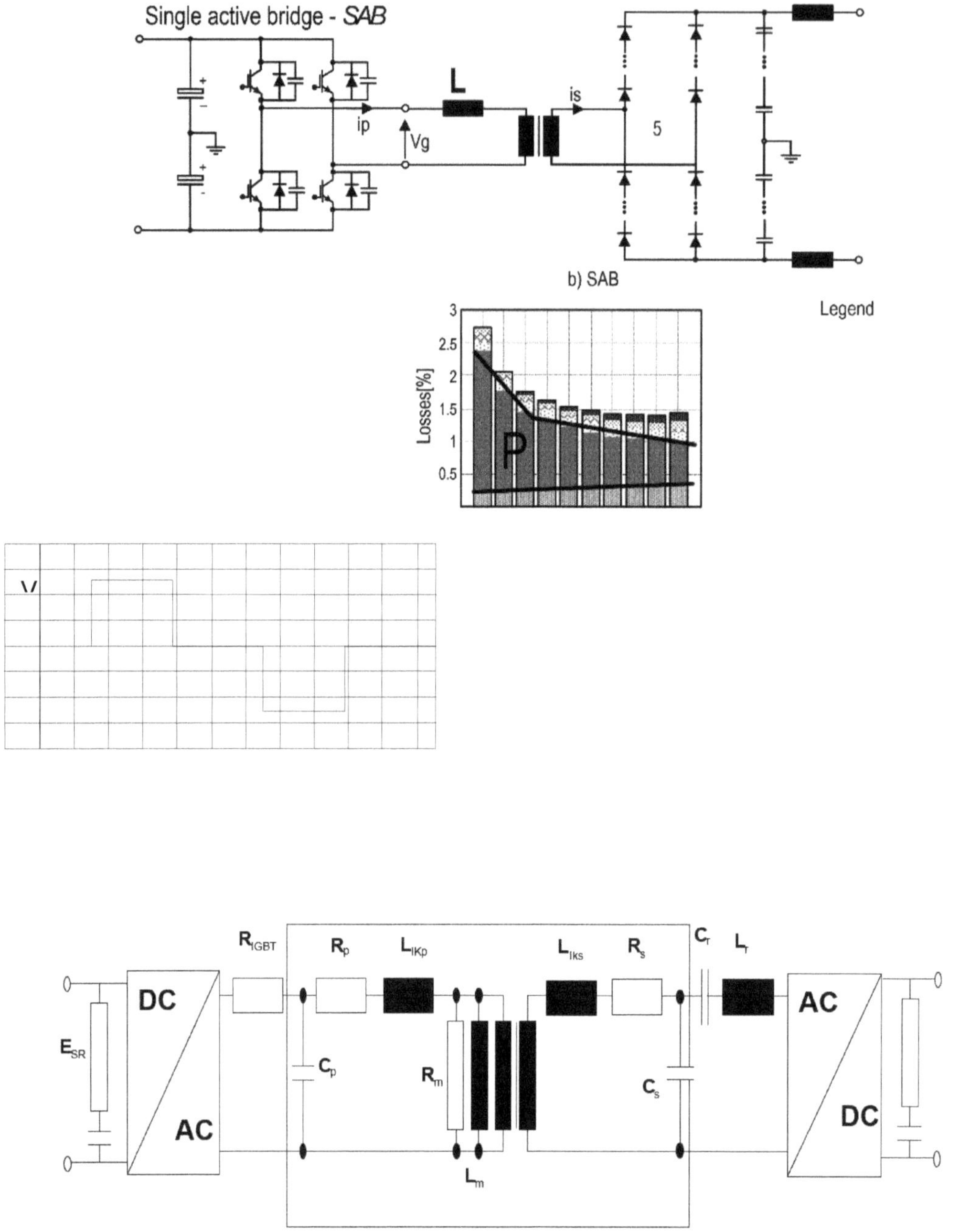

Figure 8.20 Impact of transformer non-idealities

Switching on and off the IGBTs and IGCTs is a limitation, and further, at high frequencies, the transformer non-ideal factors, like part of the resonant tank, come into play besides the Barkhausen noise of the core structure as well. Besides, at high frequencies, the stray inductances and capacitances also play a role in the smooth operation of the transformer with an added effect on stray losses. Also, it is impractical to design transformers with low leakage reactance of 0.1 for mega-watt capacity transformers. The over voltage conditions due to faults will also be attributed to tougher conditions of design for the rectifier components like the IGBTs or IGCTs. The next choice is the LLC converter, as shown below in Figure 8.21:

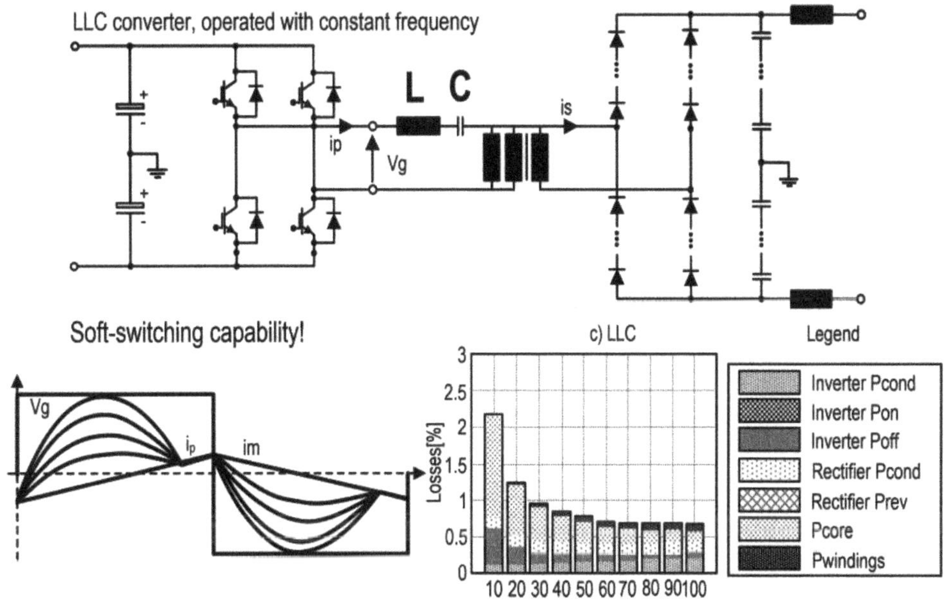

Figure 8.21 Open Loop – LLC Converter with constant frequency operation

This LLC converter operates at resonant frequency and at a frequency lower than that of the tank. This converter leads to a sinusoidal current with low turn-off current which translates into low losses, thus improving efficiency further. But, the problem is to operate at constant frequency which is going to affect the control of the LC converter. So, the question is how to control a high-power medium voltage resonant LLC converter as shown below in Figure 8.22:

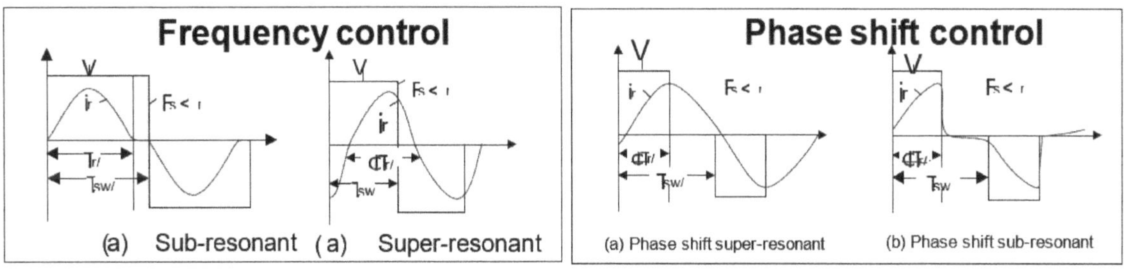

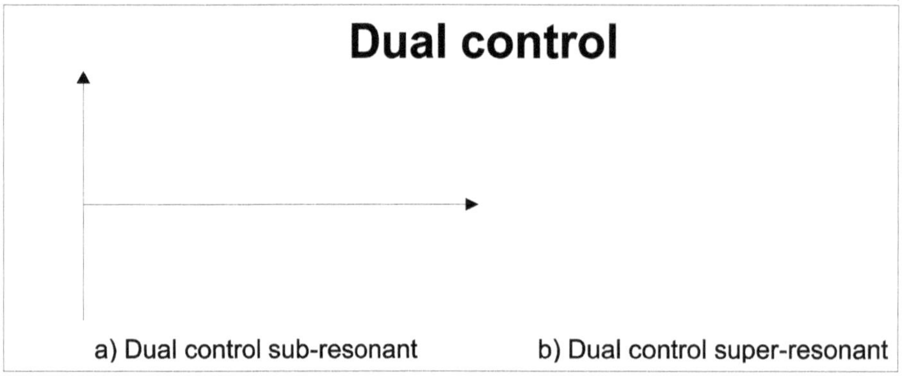

Figure 8.22 Different kinds of control employed as Frequency control, Phase Shift Control and Dual Control One can employ Frequency control, Phase shift control, and Dual control. In low voltage variable frequency, resonant converters are employed. In the HV case, attempts are made to start with a variable frequency series resonant converter, as shown below in Figure 8.23:

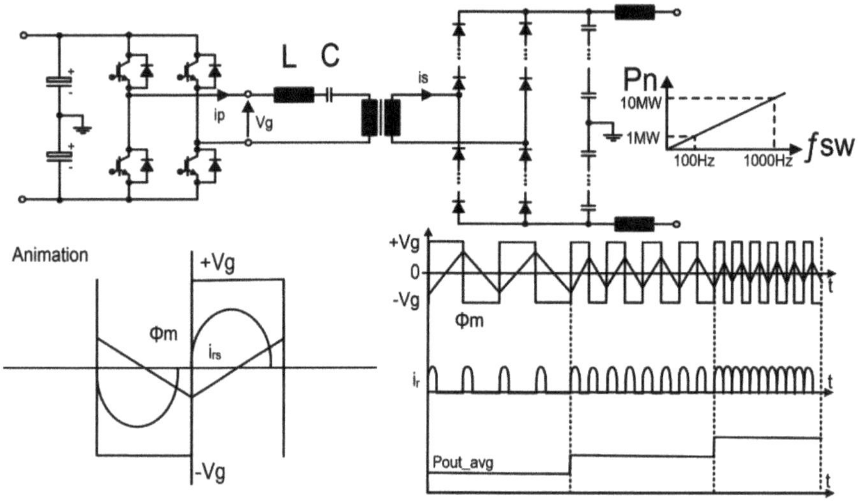

Figure 8.23 Variable Frequency Resonant Converter

The changes are shown in the next picture with variable frequency condition above as in Figure 8.24 below:

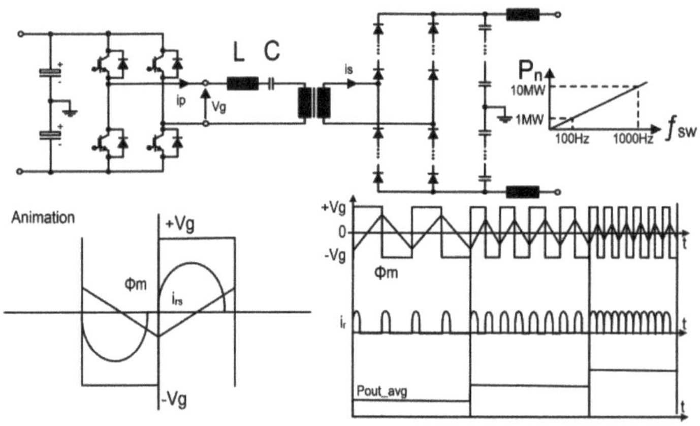

Figure 8.24 Variable Frequency Resonant Converter Performance

The LC tank filter is shown, and by varying the frequency, one can control the output power. Thus, at 1 MHz, the variable frequency can generate large power, and at low frequency, a lower amount of power as needed. The transformer has to operate at a lower frequency, so the LC tank is moved to the secondary side, as shown below in Figure 8.25:

Showing operation at two different times as below in Figure 8.25:

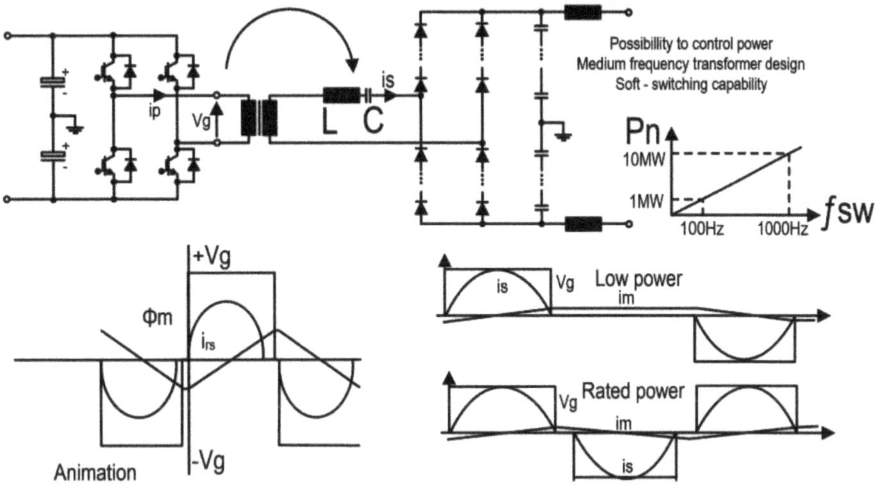

Figure 8.25 Series resonant converter

Thus, we can design a high-frequency transformer and operate it at low frequency, and the way to do it is to clamp the inverter output voltage at zero when the current also reaches zero, in such a case, we can control output power by controlling the width of pulses. And finally, the output power depends on the frequency of operation. But, variable frequency is not appreciated by the monolithic transformer as we are playing with the flux linkage. Besides, the transformer has to be designed for the lowest frequency of operation, which is typically 50 or 60 Hz. For this, the LC tank is moved to the secondary side as shown above. We can now compare the losses of these two systems as shown below, one with the LC tank in the primary and the other with the LC tank in the secondary side. Losses in the latter case will be less than 0.1%. For this reason, we selected the latter design as our choice for the proposed system, as shown below, and achieved using most of the shelf components.

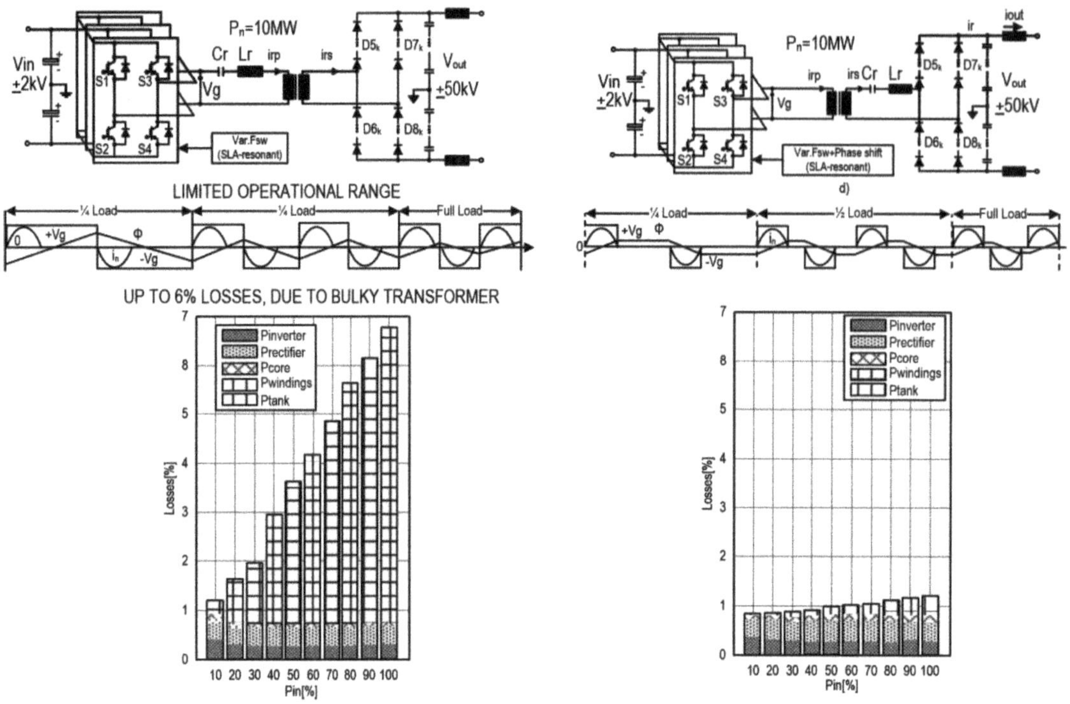

Figure 8.26 Classic SRC vs SRC#

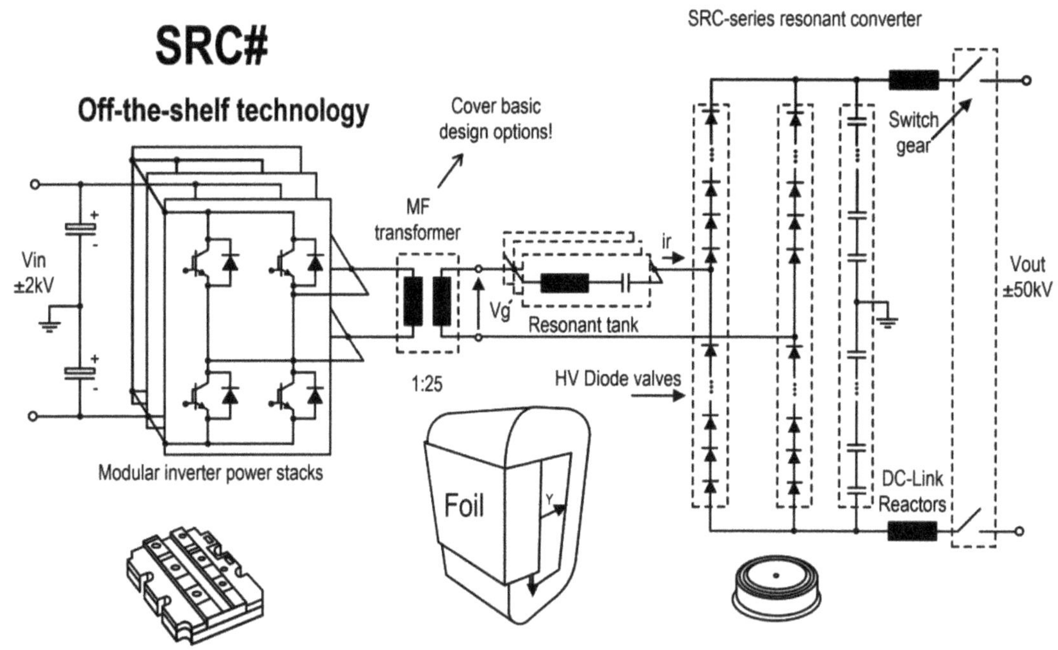

Figure 8.27 SRC# configuration

This is achieved by using modular circuit components. Placing the SRC LC circuit on the secondary side and off-the-shelf components to achieve performance, as shown below in Figure 8.28.

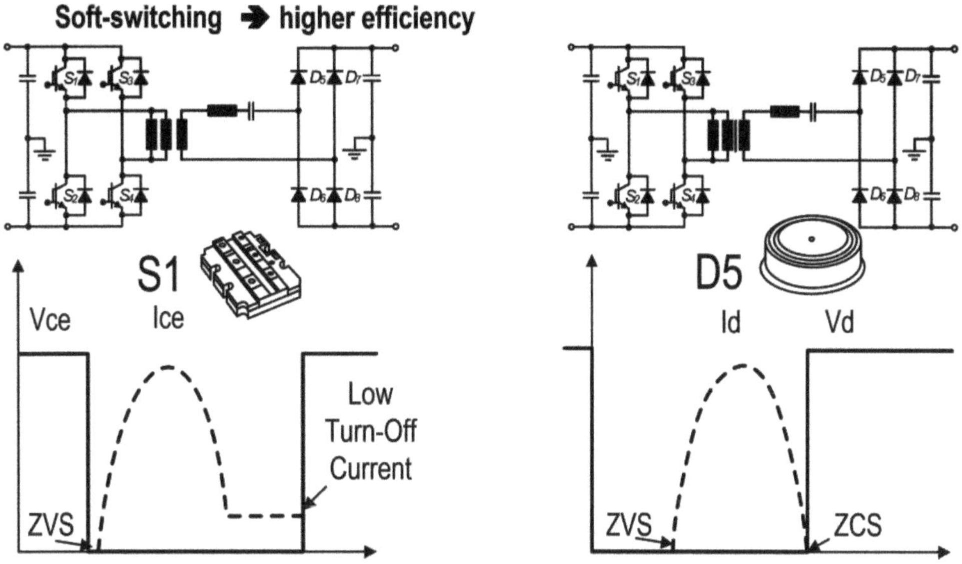

Figure 8.28 Soft switching at higher frequency

This system has soft switching and higher efficiency. It also has a low turn-off current that helps to reduce losses and then increases efficiency. Figure 8.28 also shows a comparison between circuits using IGBTs vs circuits using diodes. The converter will exhibit the following performance, as shown in Figure 8.29 below:

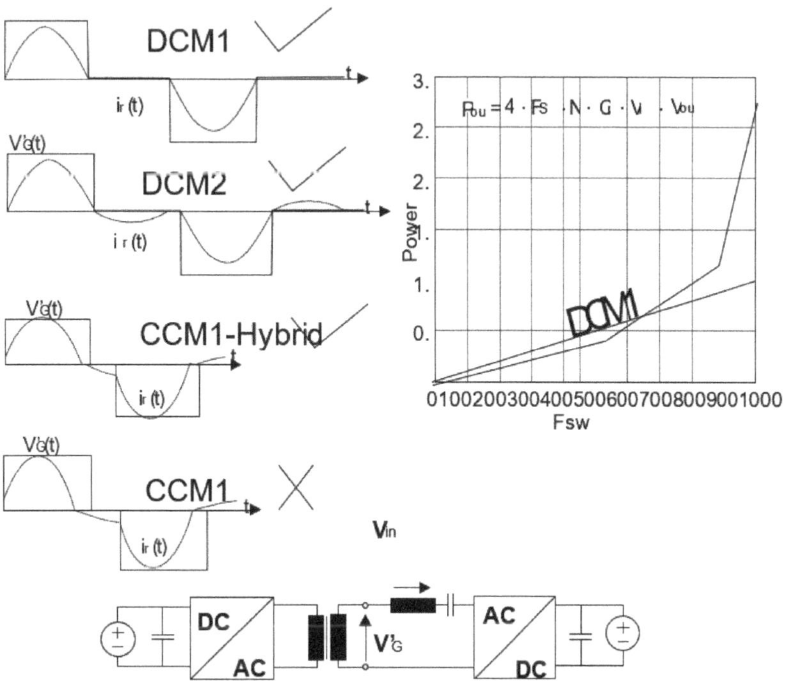

Figure 8.29 Performance of SRC#

In different modes of operation, this is exhibited, which is also dependent on the difference in voltage between the primary and secondary sides and, the frequency of operation and also the variation in frequency. It shows that in DCM mode, the operation vs frequency is linear with respect to frequency and in the CCM mode, the operation becomes non-linear. One needs to operate the converter in the DCM1 mode i.e., DC continuous conduction mode. This leads to losses of less than 1.5%. The CCM (Continuous conduction mode) mode of operation leads to non-linearity in the performance of the converter and is not the desirable configuration of operation. This undesirable element of operation non-linearity is prominent when it is difficult to control the converter, but it can be cured if there are large Voltage drops at the output.

We will assess the capacity of the diode at medium voltage in the in-line rectifier circuit, and it is found That the line frequency diode is low cost, cheap, robust, and has low conduction losses in the medium Voltage sharing region of operation.

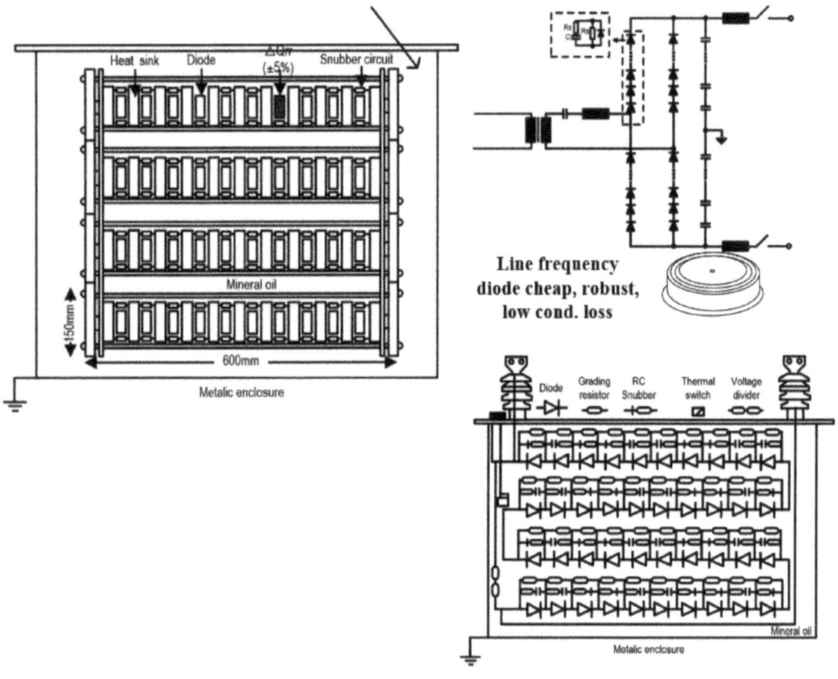

Figure 8.30 Voltage sharing in medium voltage diode

Operation details are shown below in Figure 8.31:

The operations are illustrated for +-5% and +-10% voltage variation. In the first case, the output variation is +- 45%, and in the second case it is +-10%. This is achieved by using a series of resistance and capacitance combinations, either series or parallel, but preferably with capacitor in parallel with the load.

Line frequency diode - cheap, robust, low cond. loss

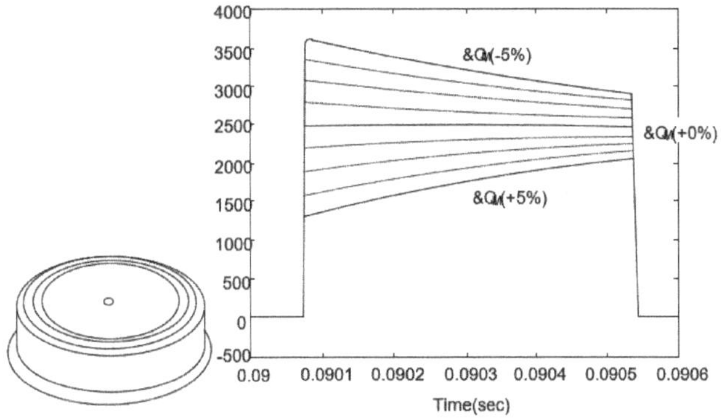

Reverse recovery charge variation

± 5%ΔQrr => ± 45%ΔV

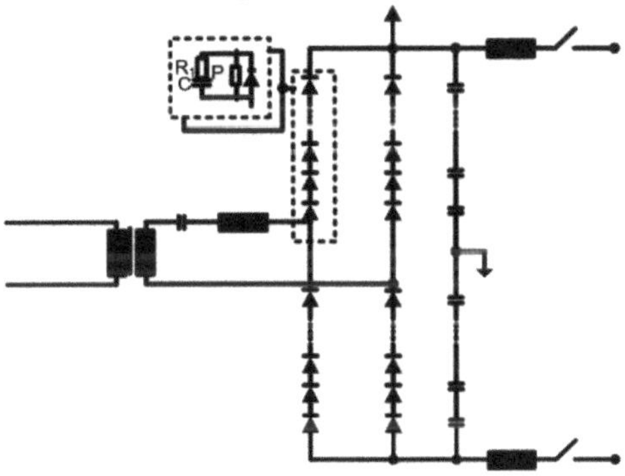

± 5%ΔCs => ± 10%ΔV

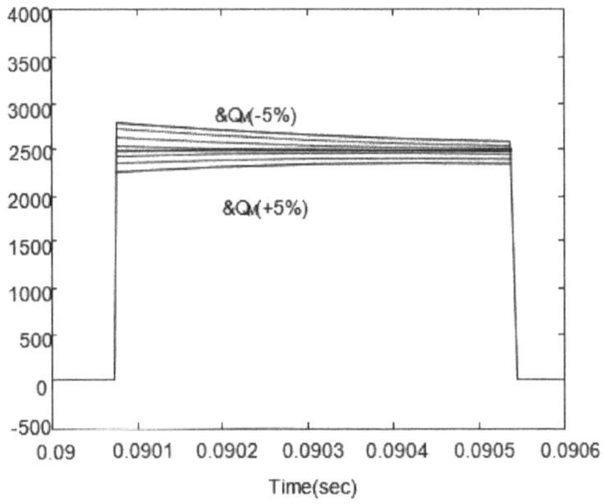

- **snubber capacitor variation**

Figure 8.31 Operation of SRC#

The performance of the topology is shown in the figure below in Figure 8.32:

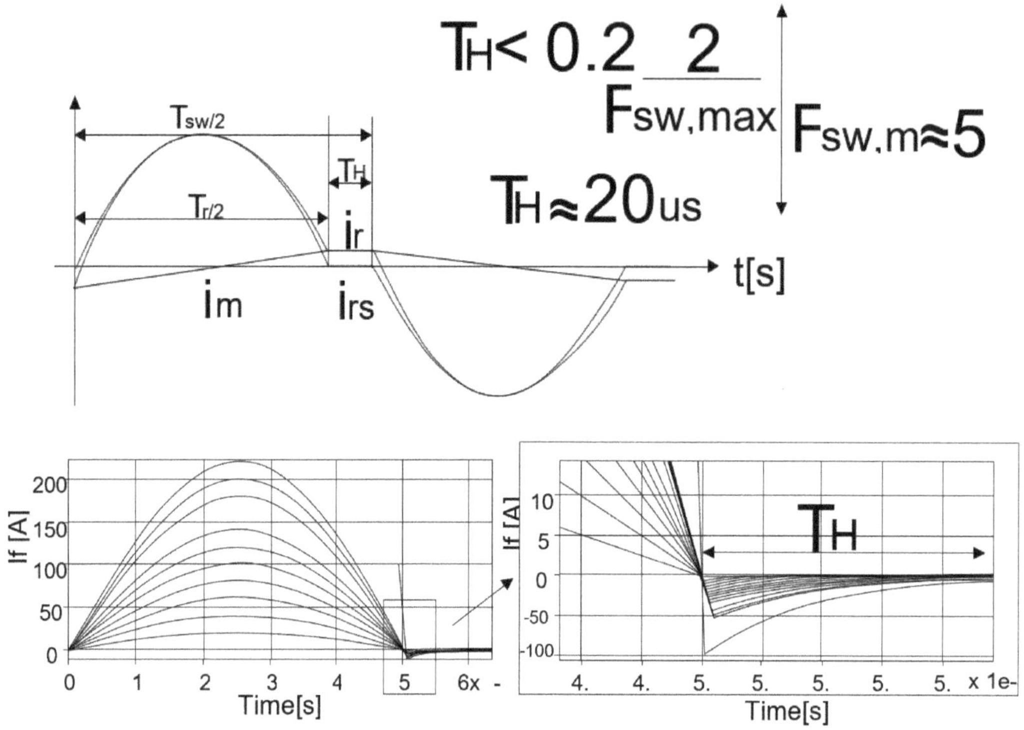

Figure 8.32 Performance of the topology SRC#

The line frequency is 5 MHz, and it shows the threshold at which the line frequency is suitable for best operation. The controller architecture is shown below in Figure 8.33.

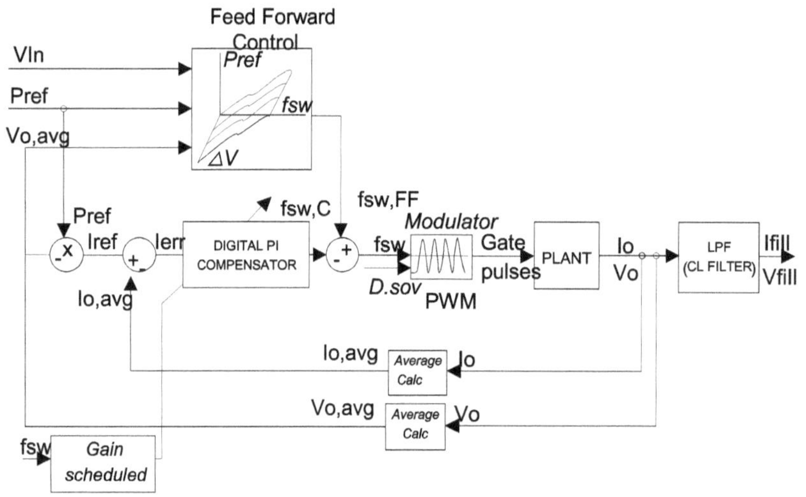

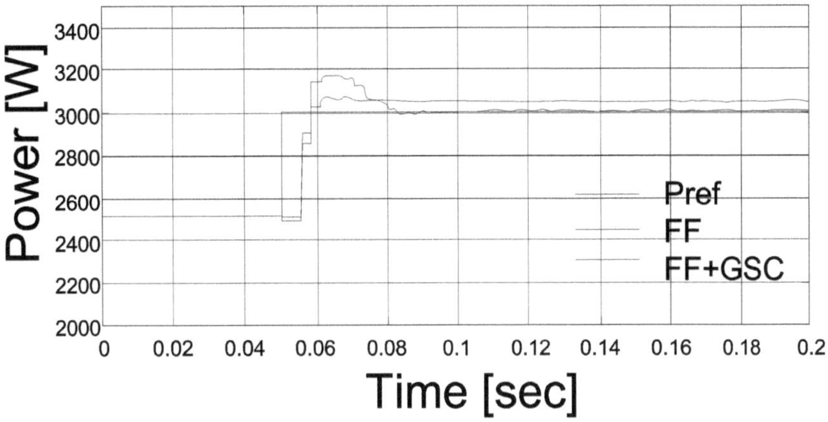

"Control aspects for a high power, resonant converter for DC wind turbines," A. Tonellotto, E. Sarra, C. Dincan, P.C. Kjaer, S.M. Nielsen, C.L.Bak, Trans. In Power Electronics and Applic., *under review*.

Figure 8.33 Performance of Controller Architecture for SRC#

The converter is designed with a feedforward controller in steady state condition. The feedforward controller is designed with the frequency of operation and output voltage variation under consideration.

A PI controller is used to control the variations in the output voltage due to the RC components in the circuit. The comparisons are between the feedforward controller w/o any feedback control, and the other is with the feedforward controller with PI feedback, which shows lower error in steady state. The best way to design such circuit controllers for the output voltage is to add parallel RC circuits to achieve the corresponding +- variations, which can be as large as +-45% if required as below in Figure 8.34.

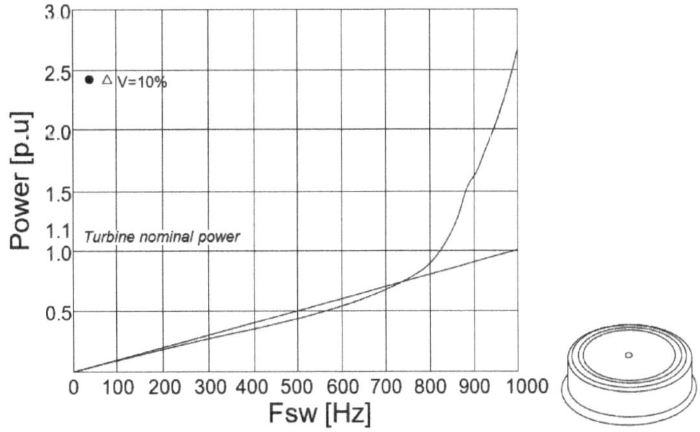

Line frequency diode - cheap, robust, low cond. loss

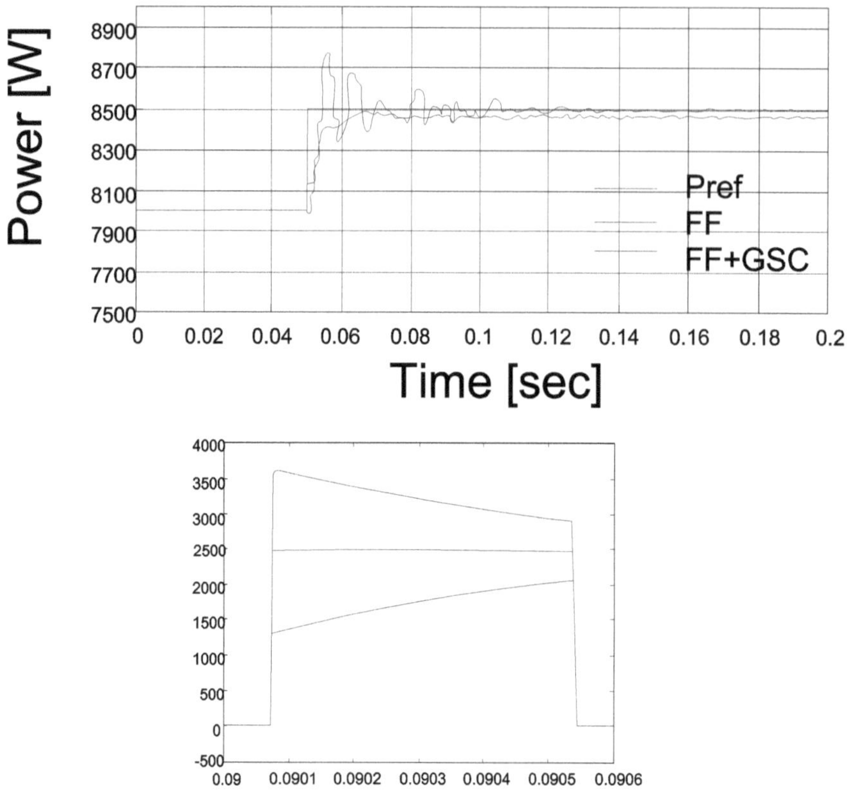

''Characterization of diode valve in medium voltage dc/dc converter for wind turbines", C. Dincan, P.C.Kjaer, Proc. IEEE Eur. Conf. Pow. Electron. And Appl. (EPE2016) September, 2016

Figure 8.34 Performance of medium voltage diode valve.

The +- 10% variations are similarly achieved by providing parallel RC circuits. The semiconductor and cooling technologies are shown in the figure 8.35 below:

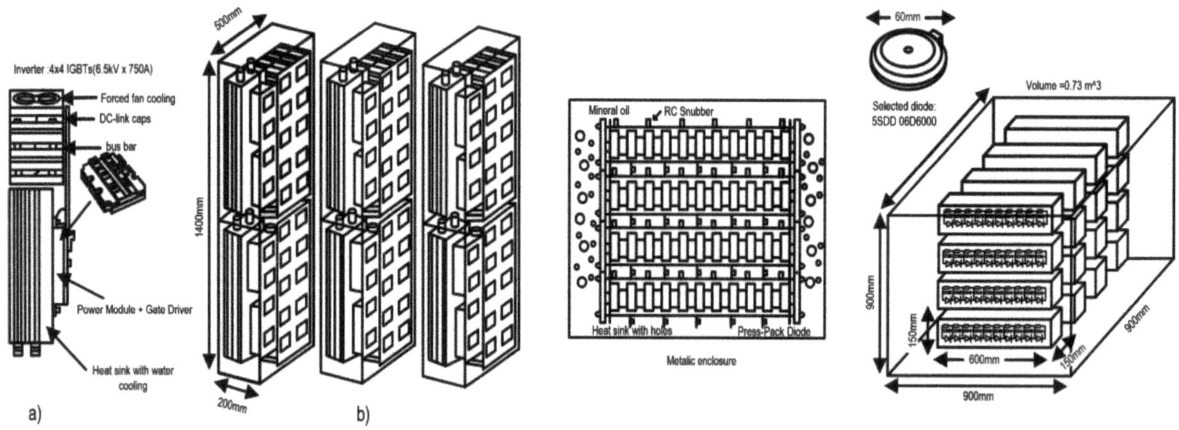

Device	650 V x 750 (FZ750R65KE3)
Number of	3x4
Total	0.2%

Device	650V x 750 (SDD06D6000)
Number of	4x4
Total	0.4%

Figure 8.35 Semiconductor and cooling technologies

The conceptual design and drawing of the medium frequency transformer are shown Figure 8.36 below:

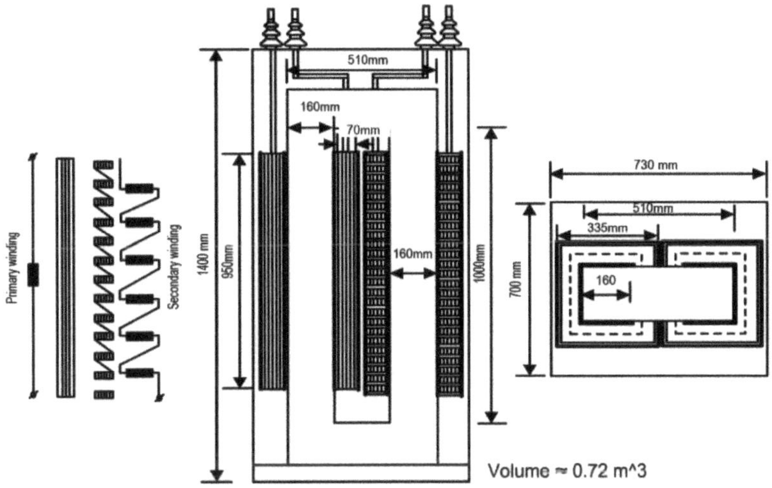

Core material	Amorphous
Flux density	1.5 T
Core material	Metglas 2605SA1
Turns ratio	25
Mag. Inductance	10 mH
Sec. Leak. Inductance	78 mH
Core mass	800 kg
Windings mass	380 kg
Total losses	0.2%

Figure 8.36 Medium Frequency Transformer, size difference between 50/60 Hz and medium frequency.

This uses the amorphous core material so that one can save on the oil content and reduce losses considerably. The line frequency diode performance is also shown in Figure 8.37 below:

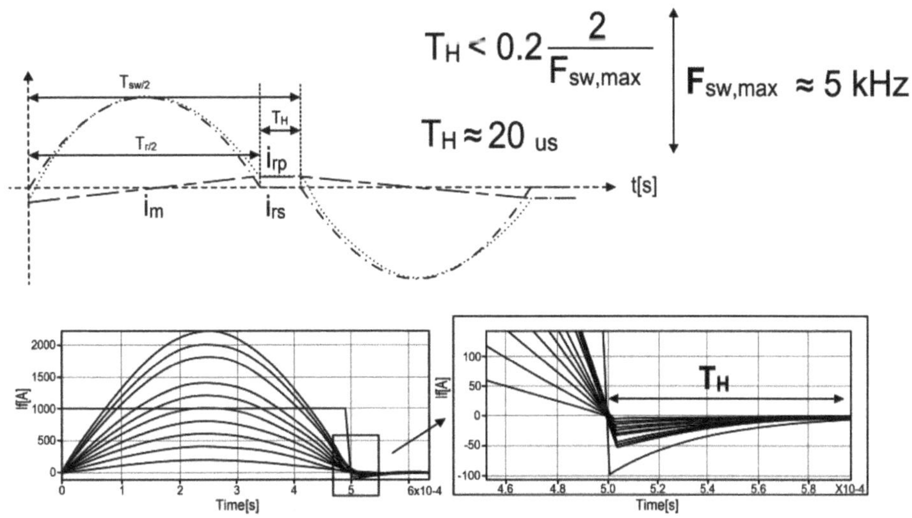

Figure 8.37 Topology of line frequency diodes

The frequency of operation is 5 KHz, and the number comes from the controller architecture discussed before and is also related to the Controller design as a feedforward controller and with PI feedback control where if the output voltage drops rapidly, the system will depict high non-linearity and in which case, a PI compensator yields a steady state operation with minimal variation, while without a Pi compensator, the performance shows a considerable variation in output voltage over a long period of time.

We will now discuss the design of 10 MW system as shown in Figure 8.38 below:

Symbol	Parameter	Value	Unit
Pin	-Nominal power	10.0	MW
$P_{in, max}$	-Nominal max. acc. power	11.0	MW
V_{in}	-Input DC Voltage	+2.0	kV
$V_{in, max}$	-Input max. acc. DC Voltage	+2.2	kV
$V_{in, min}$	-Input min. acc. DC Voltage	+1.8	kV
V_{out}	-Output DC Voltage	+50.0	kV
$V_{out, max}$	-Output max. acc. DC Voltage	+55.0	kV
$V_{out, min}$	-Output min. acc. DC Voltage	+45.0	kV
n	-Min. acc. efficiency	98.5	%p
	-Min. acc. power density	10.0	kW/L

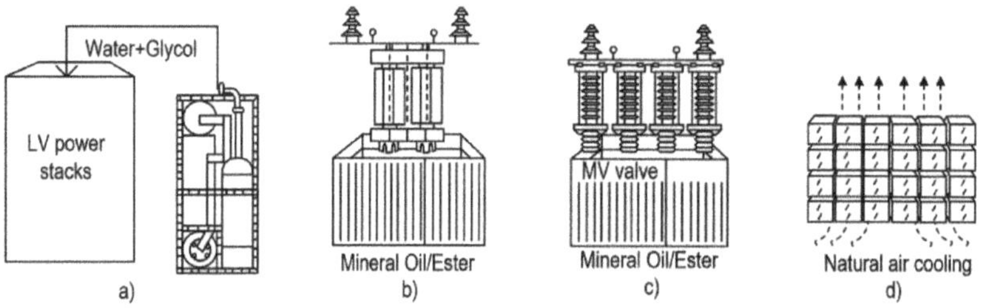

Figure 8.38 Design of a 10 MW transformer design

Here, input DC Voltage is +-2.2 Kv on a nominal value of 20 KV and other parameters as shown in Figure 8.39 below.

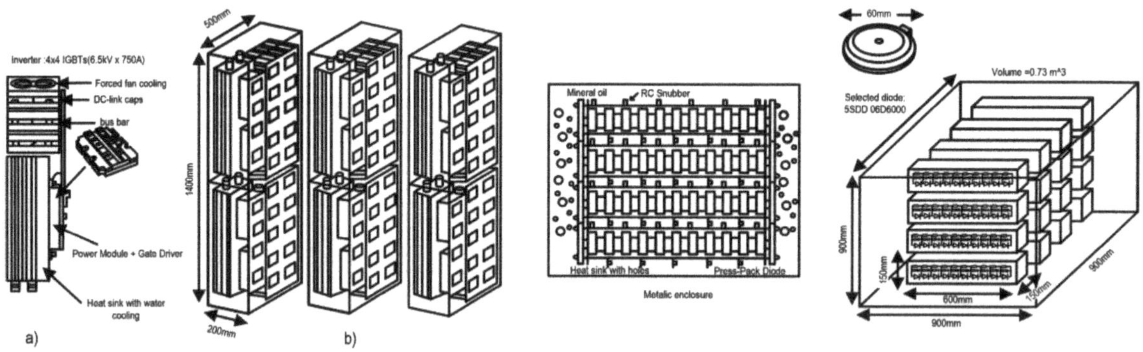

Devic	650 V x 750 (FZ750R65KE3)	Device	650 V x 750 (SDD 06D6000)
Number of	3x4	Number of	4x4
Total	0.27%	Total	0.4%

Figure 8.39 Semiconductor and cooling technologies.

There are three parallel stacks of inverter for a total of 12. The performance is described above with total 120 devices of rectifiers, total loss of less than 0.27% on the inverter side and less than 0.4% on the rectifier side. The medium frequency transformer is shown in Figure 8.40 below:

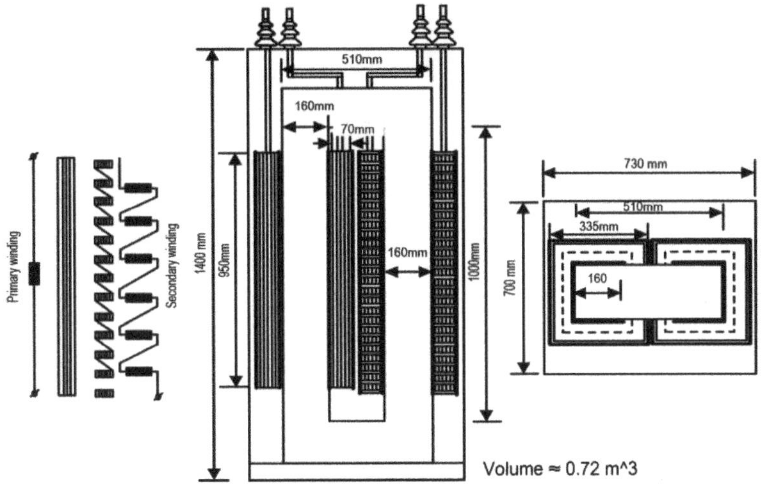

Core material	Amorphous
Flux density	1.5 T
Core material	Metglas 2605SA1
Turns ratio	25
Mag. Inductance	10 mH
Sec. Leak. Inductance	78 mH
Core mass	800 kg
Windings mass	380 kg
Total losses	0.2%

Figure 8.40 Design of Medium Frequency Transformer details with reduction in volume

The estimated volume of the transformer with an amorphous core is shown with a reduction in volume from a typical 50 Hz to 5KHz:

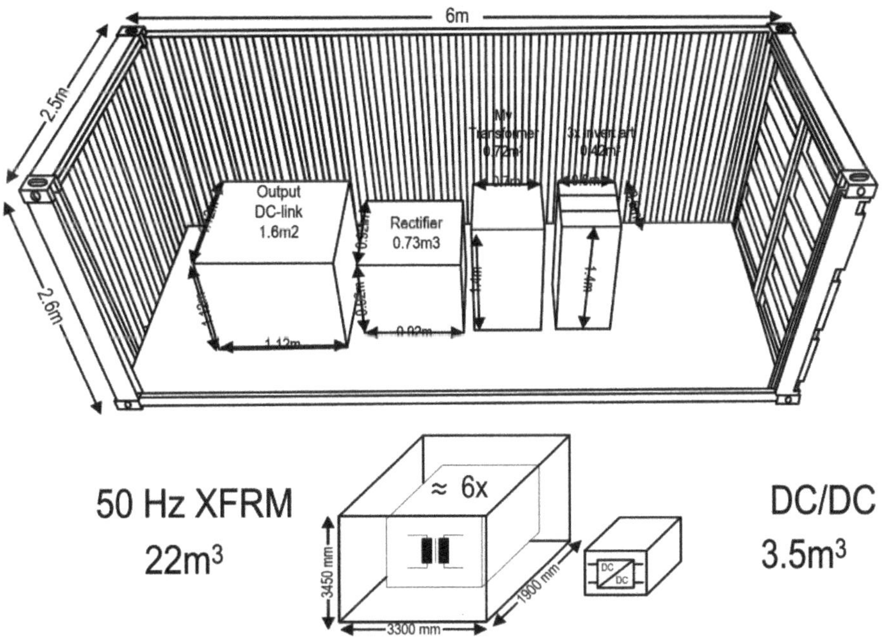

Figure 8.41 Estimated volume change between the low-frequency and medium-frequency transformer.

The DC-DC converter transformer has a reduction in volume by about six times. This saves considerably on the cost of the transformer as a capital asset. The challenges of testing at the MV level are shown in Figure 8.42 and discussed below:

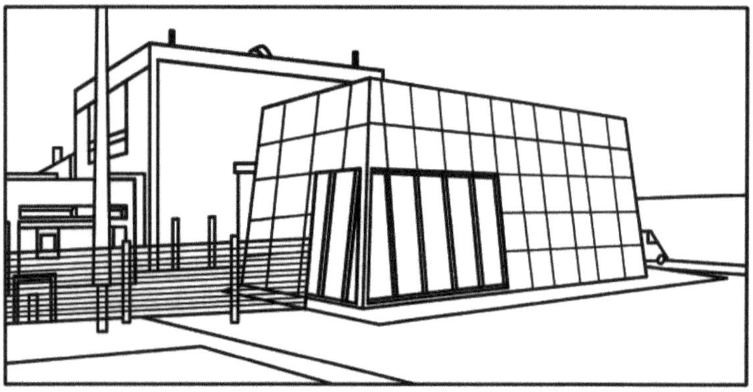

MV experiments

1. Safety issues (medium voltage testing)
2. Responsibility
3. Proper training

Testing methodology

1. Strategy of testing? – power circulation, source-sink, etc?
2. Number of iterations?
3. How to solve as much as possible with one experiment?

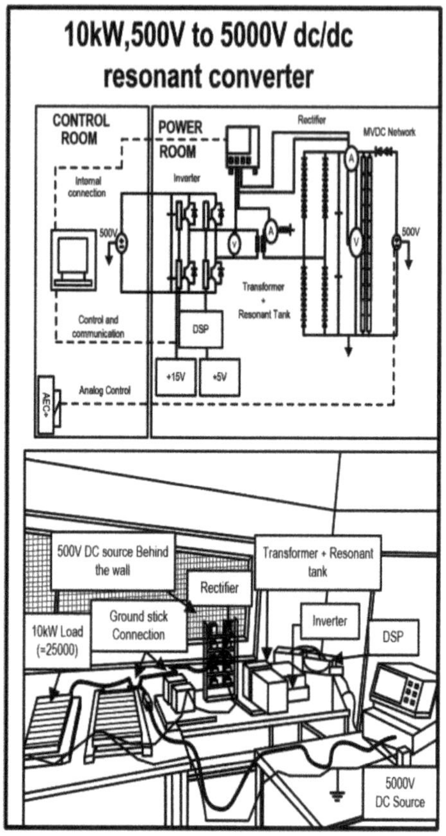

Figure 8.42 Challenges and strategy of testing at MV level.

The aspects and challenges faced during testing are mainly concerned with safety issues. The team of operators needs to sit inside the control room far away from the actual set-up of the resonant converter. Besides, proper training is necessary with all local and national codes followed stringently so as to achieve a proper functioning system at the MV level. The setup has a room that houses both the power and control room. The user sits inside the control room for safety reasons. Besides, it is imperative to formulate a proper testing methodology that has in it a proper strategy for power circulation and assessment of source and sink with respect to power system analysis. A proper valuation of a number of iterations to assess the fault levels and power flow by Newton Raphson method, clearly will help in designing an efficient DC-DC Converter system at the MV level for the Wind Turbine application. A proof of concept system is shown in Figure 8.43 below:

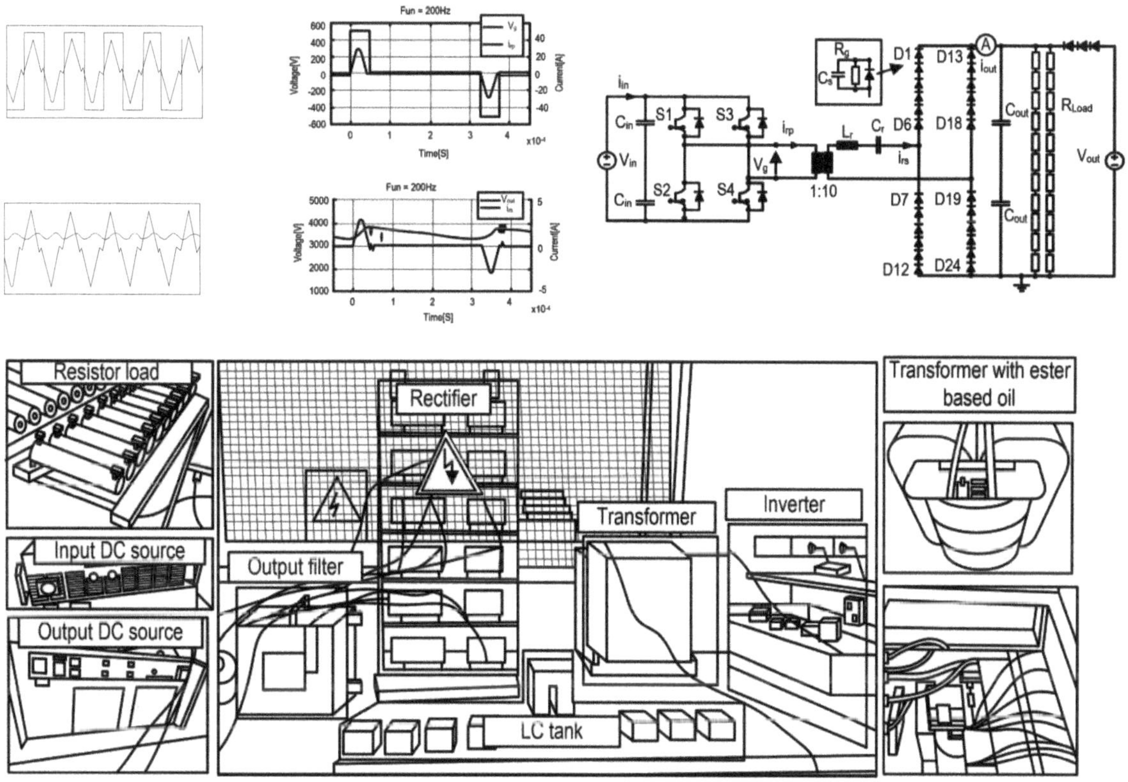

" Design of a high power, resonant converter for DC wind turbines", C. Dincan, P.C.Kjaer, Y. Chen, S.M.Nielsen, C.L.Bak, E. Sarra, V. Sriram – Trans. In Power Electronics and Applic., *under review*

Figure 8.43 Proof of concept of the system with performance

This system is for a 10 kW, 500/5000V DC-DC system with a transformer with a 1:10 ratio and having an amorphous core and designed at 5 KHz, with coil windings. The rectifier design had line voltage diodes, and the output side was the resonant frequency converter at the MV level. The result was an efficient system that can be employed for a wind turbine grid system. Shown below in Figure 8.44 are actual recorded waveforms:

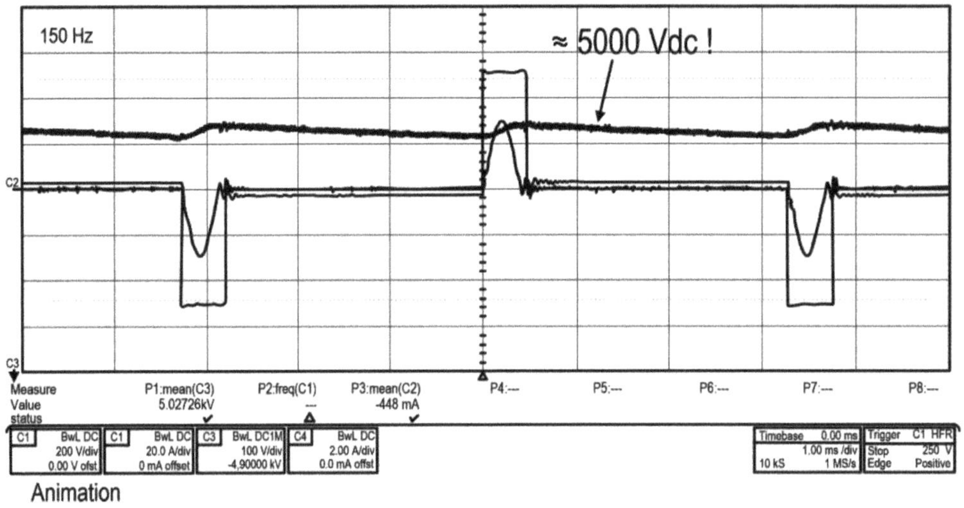

V_g – inverter output voltage V_{out} – output voltage ripple i_{rs} – secondary resonant current i_{rp} – primary resonant current
Figure 8.44 Output voltage ripple and primary resonant current.

The output voltage ripple is illustrated in the video recording as shown in Figure 8.45 below:

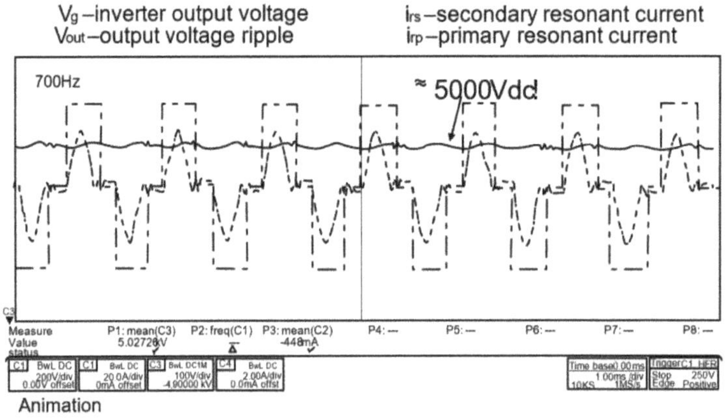

Figure 8.45 Output Voltage ripple and primary resonant current

Many salient parameters can be seen in these recordings, like the magnetizing current in the transformer, the ripple on the output, the primary resonant current, and also the secondary resonant current. The measurement of losses was achieved by setting up a testing system, as shown below, so that proper assessment of losses in the rectifier, transformer, and the converter can be easily measured with low voltage probes on the primary side and high voltage probes on the secondary side, and then the validation can be made with the help of the MATLAB algorithm for such a prototype system.

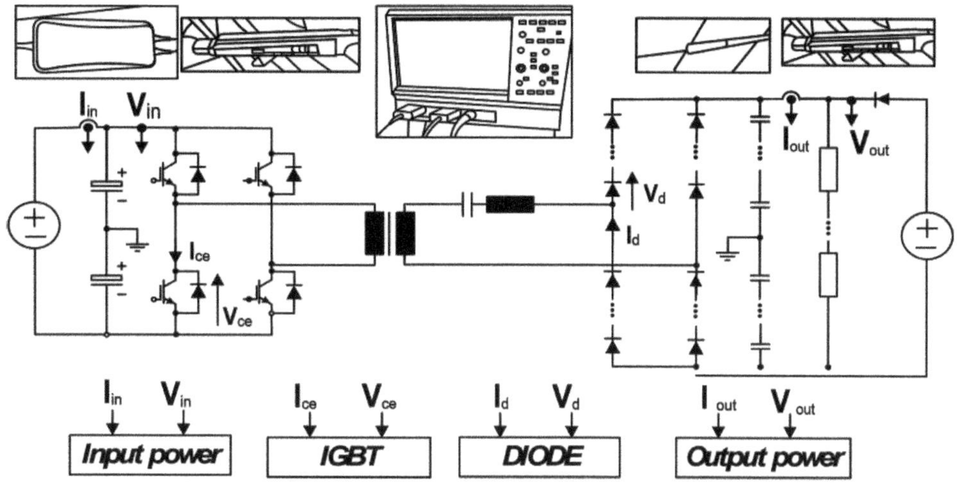

Figure 8.46 Loss measurement of the prototype system

The arrangement is as shown above. The measured losses are shown in Figure 8.47 below for the proof of concept system. The losses range from 6% to 4.5%. The inverter losses and the converter losses match with the model, and this shows a viable prototype DC system for a wind-turbine-operated grid.

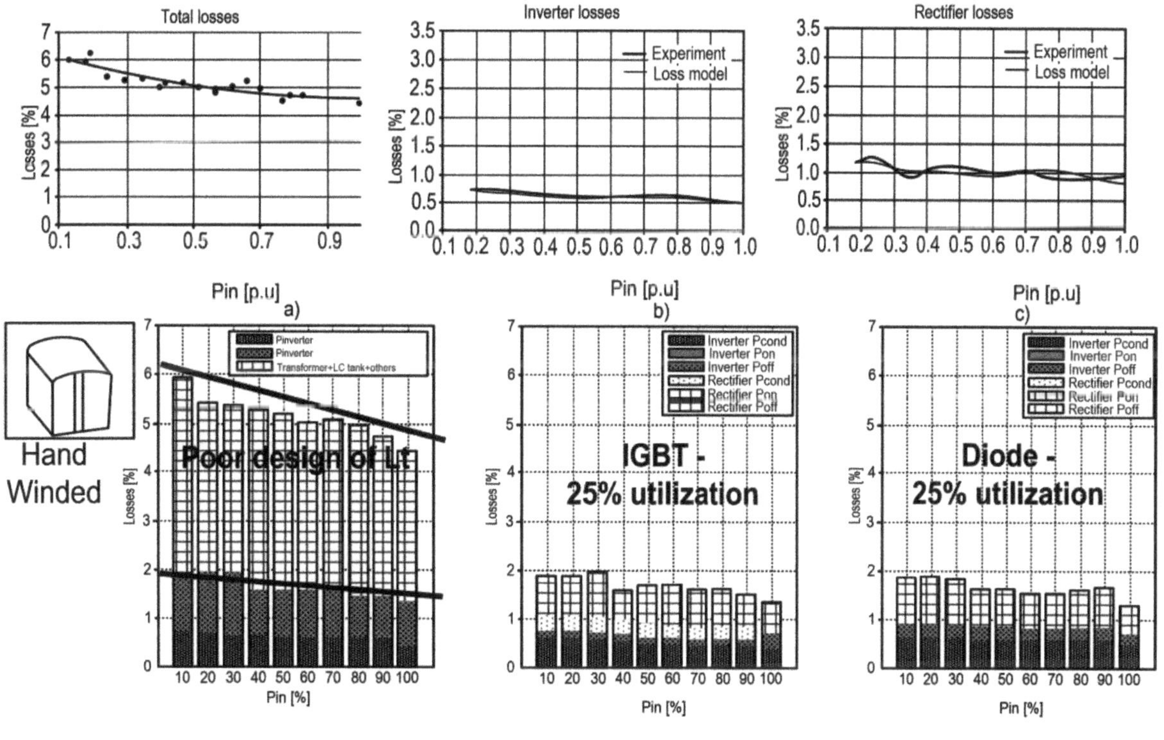

Loss segregation of measured losses (a); loss segregation on measured (b) and simulated (c) semiconductor losses. Figure 8.47

The IGBTs and Diodes were utilized for 25% and were not the optimum devices for the specifications, but that can be easily rectified by matching the requirements and performances in the future. The objective was also to find the voltage sharing in the rectifier diodes, and this is shown in Figure 8.48 below:

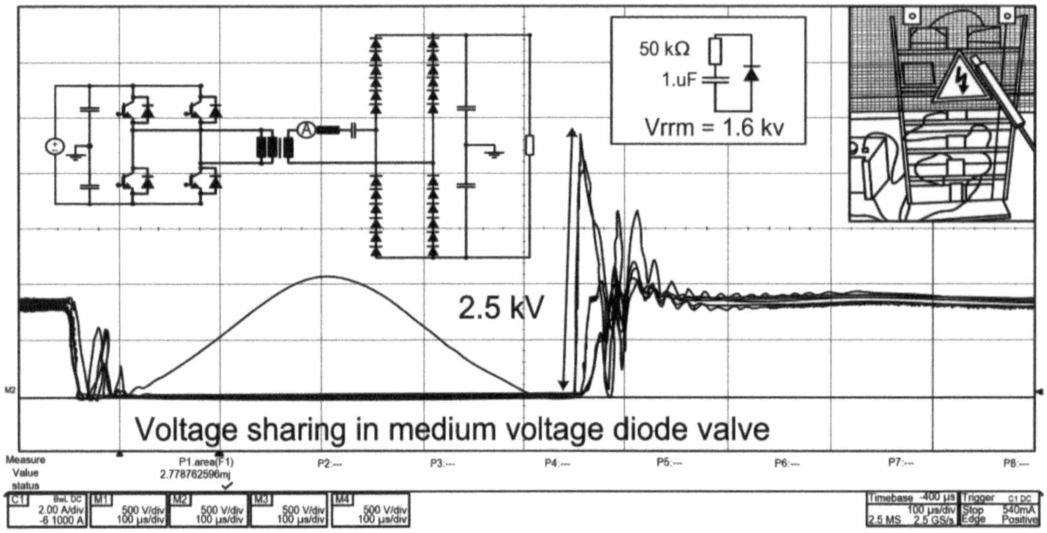

Figure 8.48 Performance of proof of concept.

It can be seen on one of the branches, and it is seen that that these diodes can block faster than the other branches. But with the proper design of the RC circuit, we can design a proper sharing as shown in Figure 8.49 below:

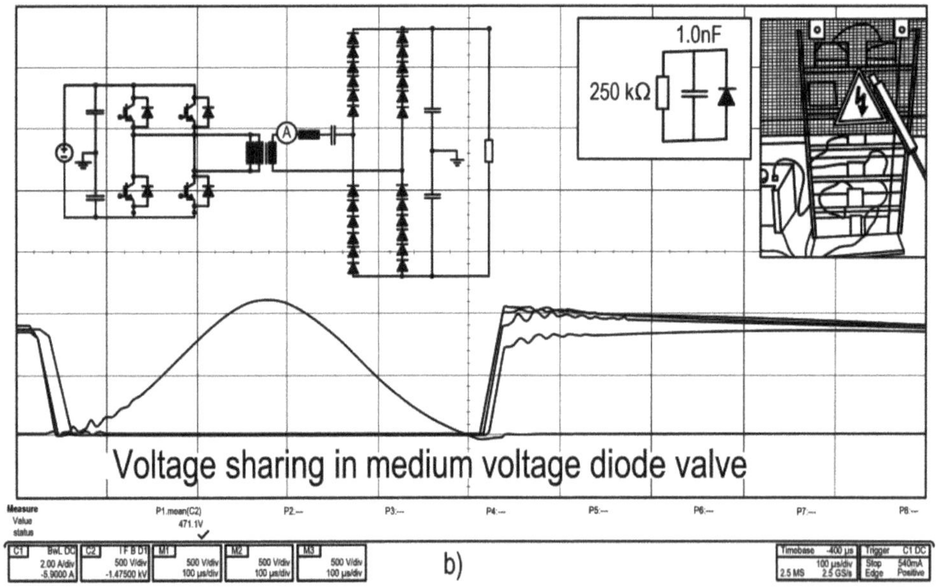

Figure 8.49 Further performance of voltage sharing in medium voltage

Further, the next exercise was to conduct an exercise to verify the soft operation of the proof of concept system as shown in Figure 8.50 below:

Vin	4000V
Vout	1000V-4000V
Cin	101 uF
Cout	106 uF
Lm	55 mH
Lr	180 uH
IGBT	6.5 kV/0.75 kA
Diodes	6.5 kV/0.75 kA

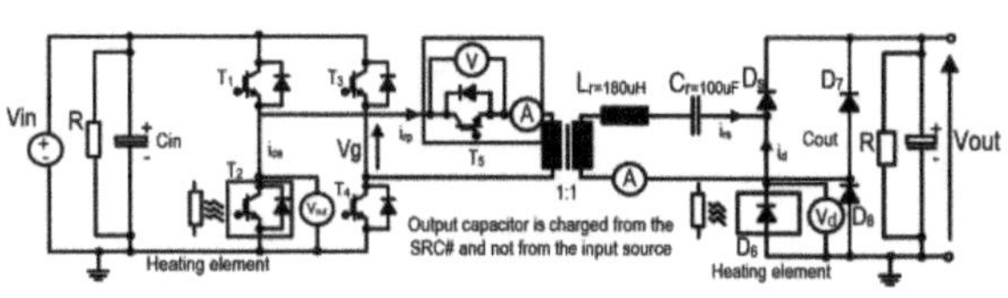

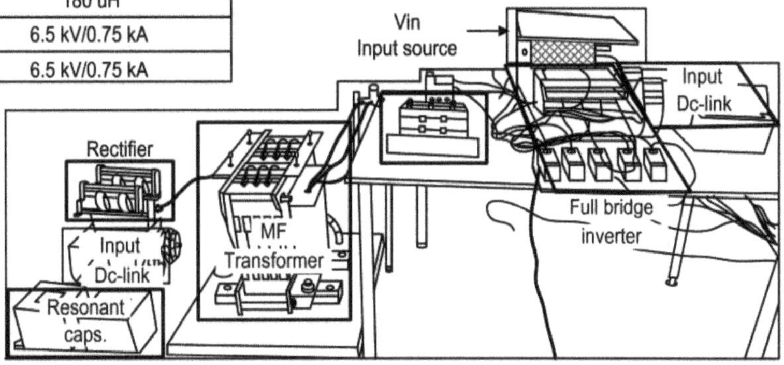

"Soft-switching characterization of medium voltage IGBT power modules and press-pack diodes in a 1kHz mega-watt dc/dc resonant converter", C. Dincan, P.C.K Jaer, S.M. Nielsen, C.L. Bak, Proc. IEEE Eur. Conf. Pow. Electron and Appl. (EPE2018)

Figure 8.50 Soft switching experiment with target devices.

The setup for the principle of operation is shown in Figure 8.51:

Figure 8.51 Principle of Operation of soft switching experiment with target devices

The output current was restricted to 2.0 amps when the gate signal was applied, and the PWM was synchronized, as shown below in Figure 8.52. Some results show good results:

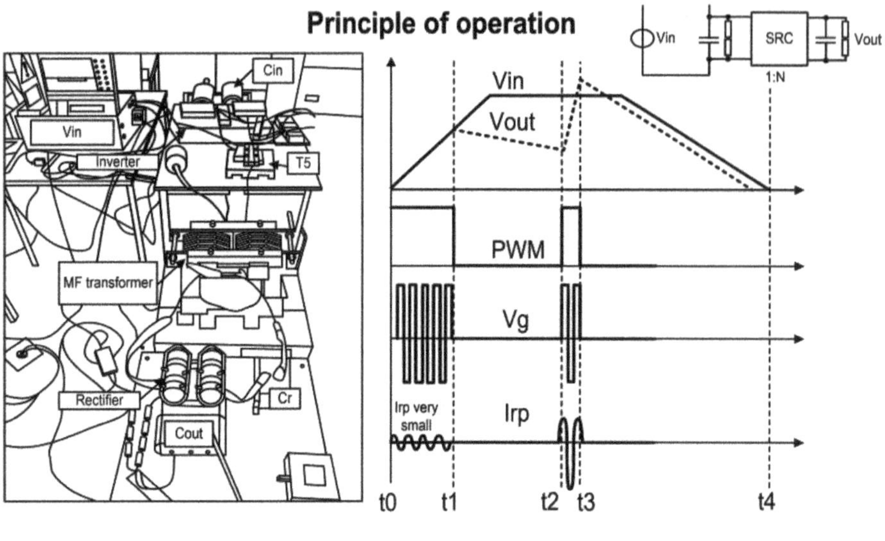

Figure 8.52 Performance of proof of concept

At 80 and 300 Hz, respectively, with voltage reaching 4000 volts and current reaching a few amps. The Currents recorded on the IGBT are shown below in Figure 8.53:

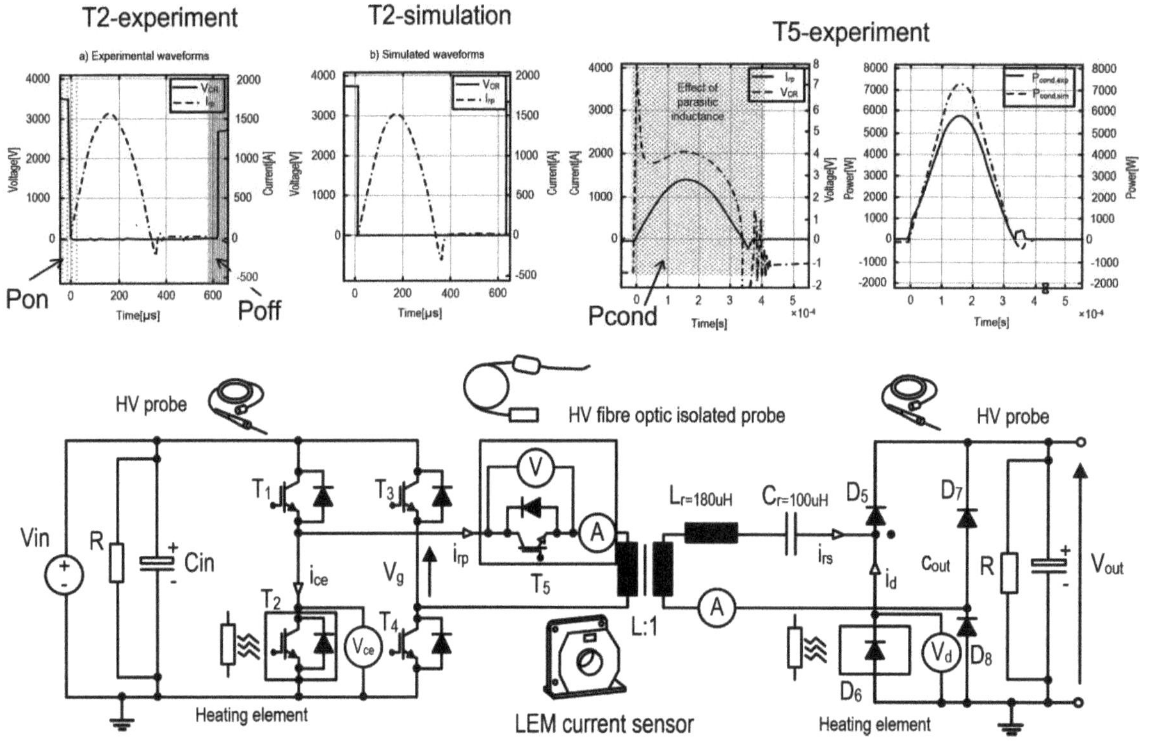

Figure 8.53 IGBT losses

Similar results can be created in the simulation model. The IGBT losses were evaluated using high-level fiber optic probe and results compare well with experiment, simulation and those measured by the fiber optic probe. The difference seen is due to the leakage inductance. The IGBT turn-on and turn-off losses are shown below in Figure 8.54:

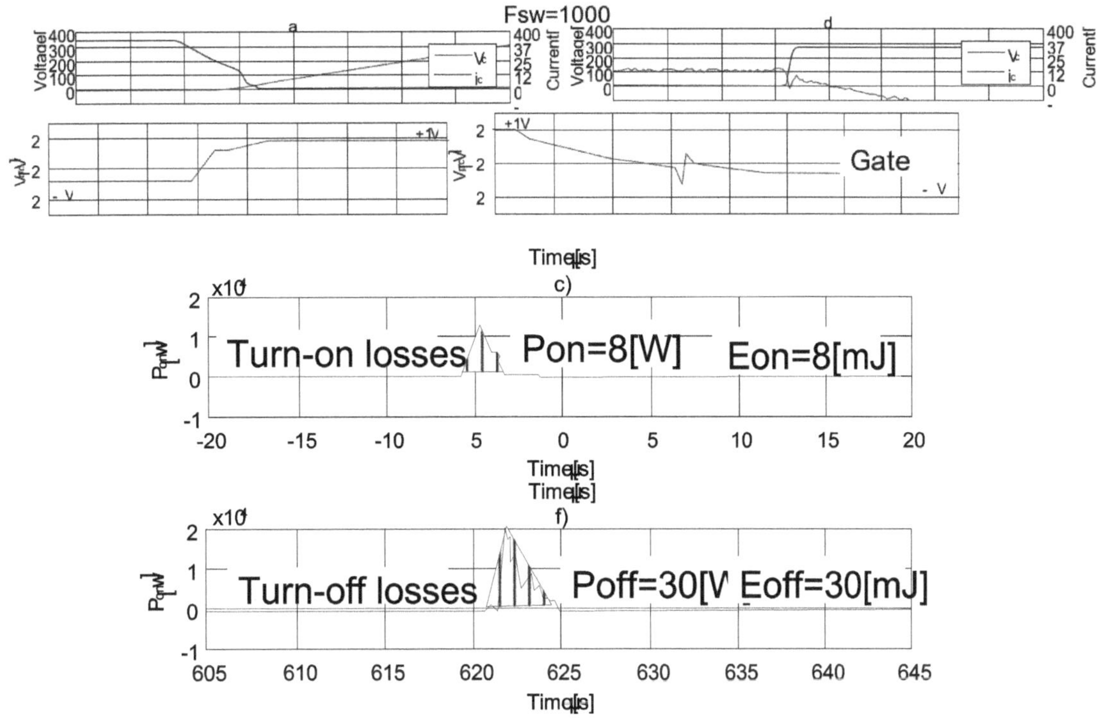

Figure 8.54 Turn on process, turn off process.

These losses are very low, and the losses should actually be zero, but some irregularities in the setup can lead to some marginal losses, as seen above in the IGBT's turn-on and turn-off process. Also the magnetizing current in the transformer is offset and does not have enough time to reach a steady state value. We also look at the diode losses as shown below in Figure 8.55:

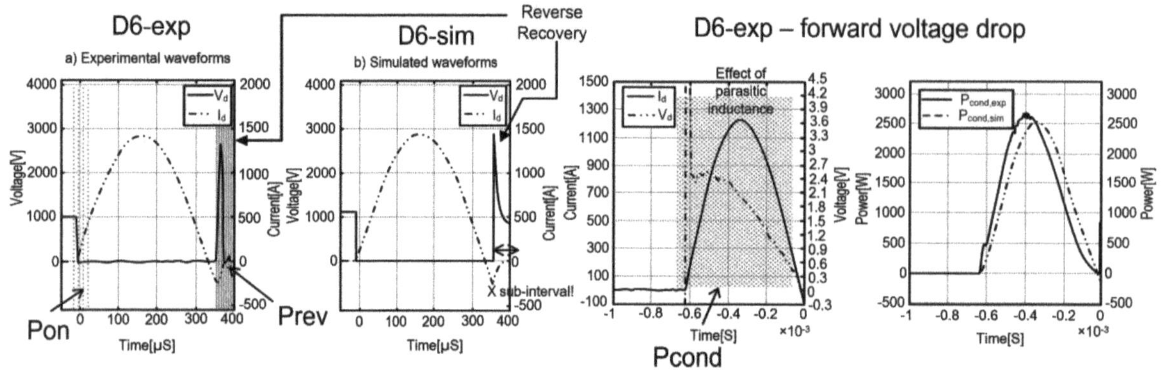

Figure 8.55 Diode losses

To observe the forward voltage, drop, the output capacitor was charged to 1000 volts and the experimental and simulated losses in the diodes are comparable. The spike seen in the voltage is due to the recovery process, and then it drops to zero, as seen above. Similarly, by zooming in, one can see the diode losses as the process continues, as below in Figure 8.56:

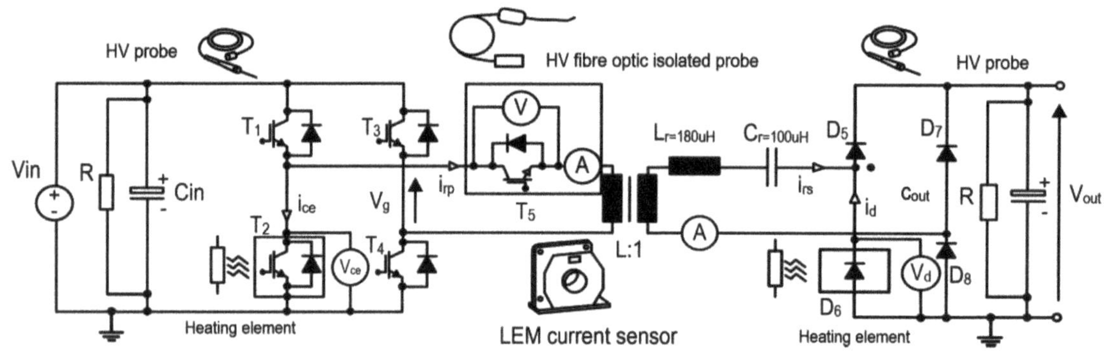

Figure 8.56 Diode losses further described

During the turn-on and turn-off processes, the losses are zero. The measured and calculated losses are shown below where there is some difference, which can be explained as follows in Figure 8.57:

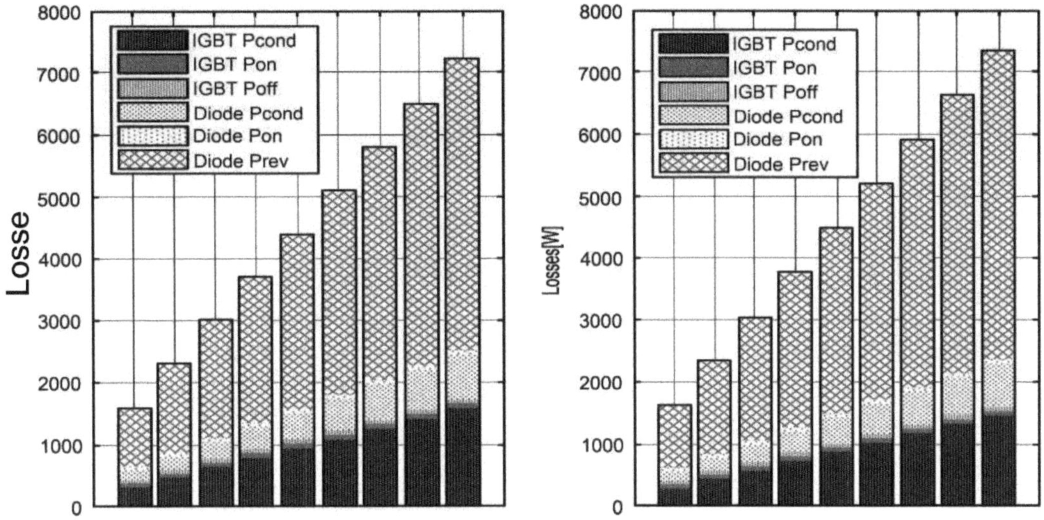

Figure 8.57 Measured losses on soft switching set-up

While the predicted losses are shown below in Figure 8.58:

Pn = 10MW

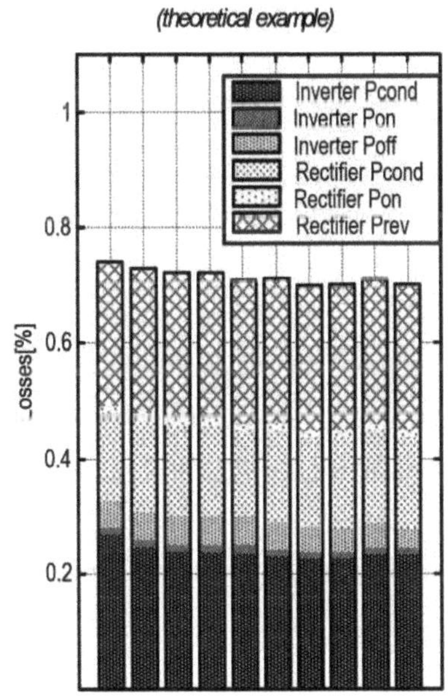

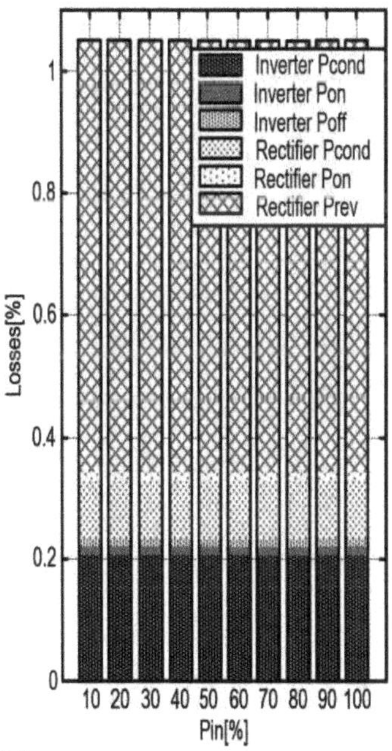

0

10 20 30 40 50 60 70 80 90 100

Pin [%]

Inverter losses **Rectifier losses**

0.27% **0.41%**

Pn ≈ 2.8MW;

average power during one pulse with Fsw = 1000 Hz

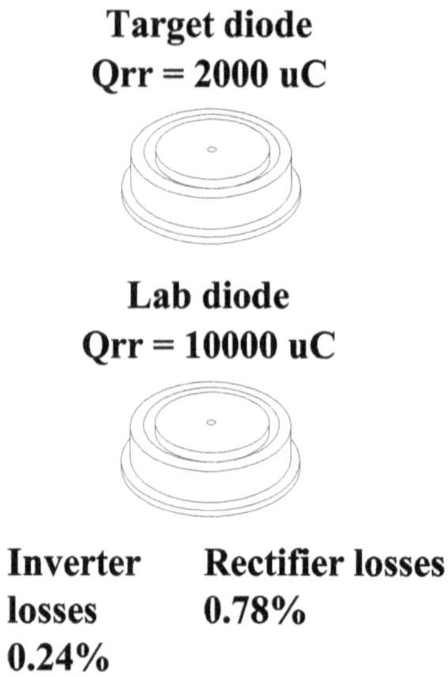

Figure 8.58 Predicted losses on 10 MW prototype

This was because of the avalanche diodes used in the lab and these were not the target diodes proposed for the proof of concept system. Thus, we can see that the level of confidence is very high in the proposed system, with expected losses for measured, simulated, and predicted being comparable.

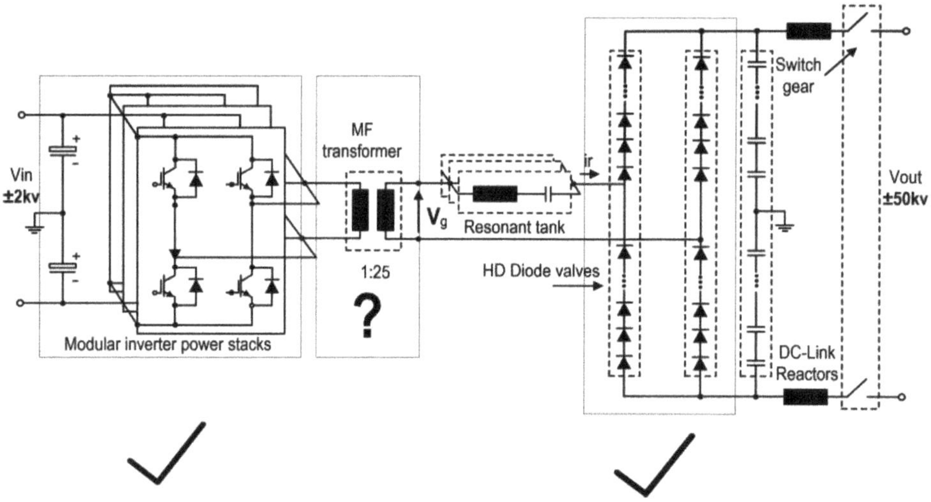

Figure 8.59 Soft switching for semiconductor model with high confidence

Thus, we can conclude in Figure 8.60 as described below:

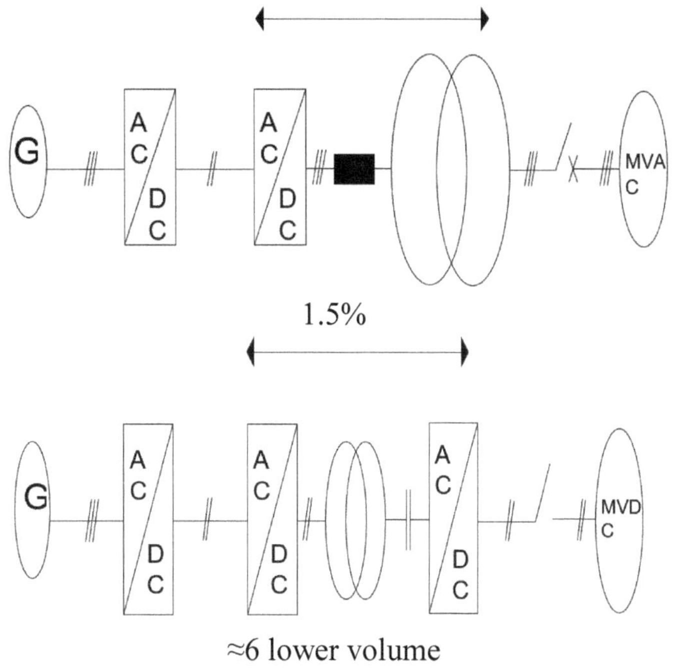

○ A high power, medium voltage, resonant DC/DC converter (SRC#) has been proposed for MVDC application ○ Medium voltage experiments have been performed and technology readiness level has been increased Figure 8.60 Summary Conclusions The main conclusions are shown below in Figure 8.61:

- **A new modulation scheme** introduced for a single-phase series resonant converter
- **A new converter philosophy** for DC wind turbine application was proposed
- **Circuit conduction modes identified**
- **Increased technology readiness level** to TRL4 with experimental proof of concept
- **Validation** of semiconductor loss model
- **Novel soft-switching characterization** setup introduced
- **A closed-loop control architecture** was proposed and validated
- **A design guideline** for the SRC# is proposed

Figure 8.61 Main Conclusions

Future work to be done is shown below in Figure 8.62:

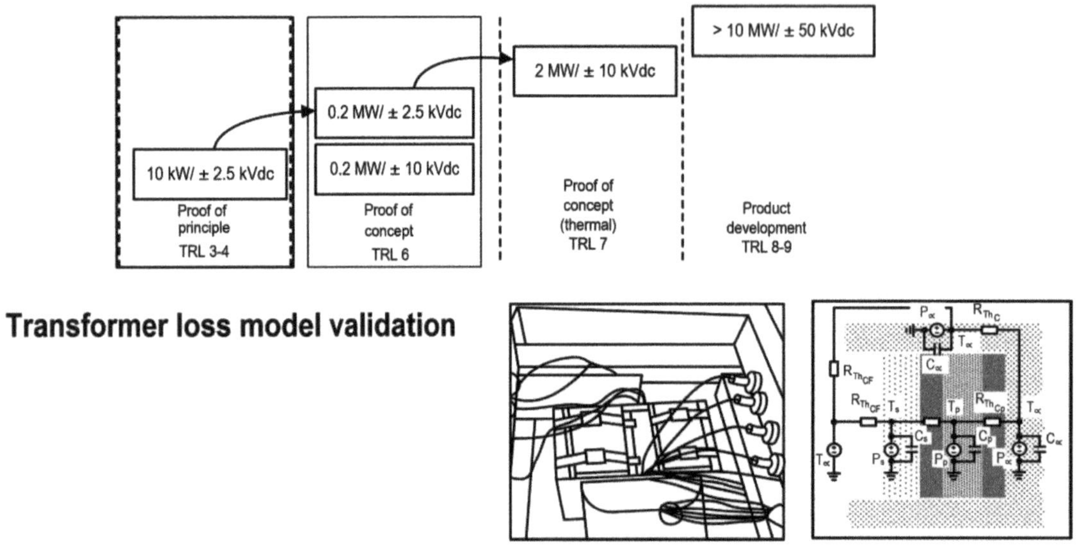

Figure 8.62 Outlook and Future Work

Transformer loss modelling needs to be done through parametric methods.

Chapter 8 Problems for Prospective System

Problem 1:

From the material presented in this chapter, what are the salient factors to be considered before a future solution is finalized as compared to the present solution for an MVAC+MVDC offshore wind farm? Please describe in detail your individual thought process with emphasis on DC-DC converters of different types to be used in your design.

Problem 2:

Conduct research and suggest a system to conduct a system as presented in this chapter.

2 HV DC-DC Converter for Electric Vehicle

Recently Eaton has developed an HV DC/DC Converter for Electric Vehicle application, which changes the higher voltage of the battery to lower voltage to power infotainment and safety systems. These converters are highly reliable and can design solutions while leveraging production processes.

Figure 2.1 HVDC DC-DC Converter for Electric Vehicle with 19 volts in and 5 volts out (courtesy client, manufacturer)

Chapter 8 Problems - HVDC case study

3 Synchronizing of multiple DC-DC Converters

Important aspect of this methodology is to synchronize the switching frequency of the multiple DC-DC converters. Below is a block diagram of two converters being synchronized?

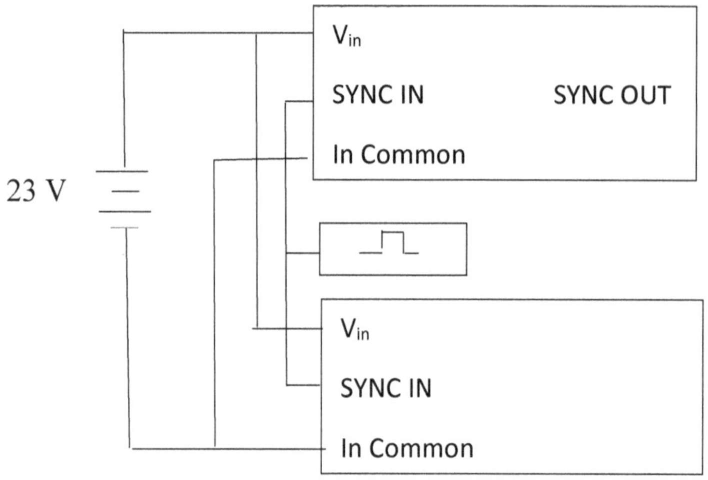

Figure 1 Multiple DC-DC Converters being synchronized of the same bus

Two converters with sync-in capability are provided with an external sync-in source to synchronize the Two DC-DC converters at the same switching frequency. In addition to this SYNC in pin, manufacturers can also provide a SYNC out pin. This can be used to sync multiple external converters with the master converter with a sync out pin. All synching is referenced to the primary ground. The synchronizing signal is common to all converters and should be TTL compatible. The synchronizing signal is a maximum of 6 Volts, as shown below, and should not be allowed to go below 0.3 volts. This may cause damage to the synchronizing signal. The duty cycle of the synchronizing signal is maintained at 20%.

There are different synchronizing techniques and types provided by different manufacturers like VPT Inc.

CASE STUDY II – Grimsel-II, Hydroelectric Substation, Innertkirchen, Switzerland

This station is located two hours south of Zurich in Switzerland. I was invited to test transformers for partial discharge emission at this plant in February of 2020.

The plant consists of four DC Generators at 25kV that can generate a total of 125 MW. This energy is then fed to the system as shown below: First, the DC voltage is inverted to AC voltage at 25 kV. The AC 3-phase voltage is then stepped up to a higher intermediate AC voltage of 145 kV. The 3-phase AC voltage is then rectified to DC Voltage at 145 kV at MVDC. A DC-DC Boost Converter then steps this up to a 525 kV DC Voltage at HVDC. This HVDC at 525 kV is then inverted to 525 kV 3-phase AC, as shown below in Figure:

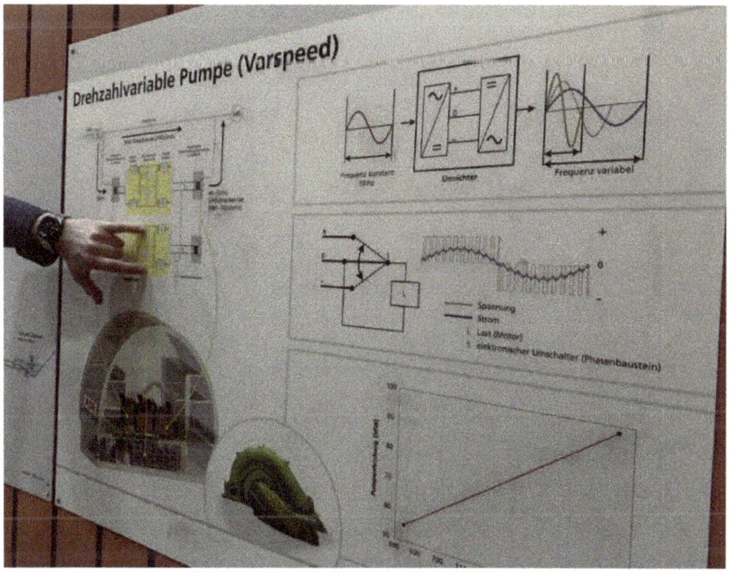

Figure 8.2.1 Pump Storage Plant (courtesy client)

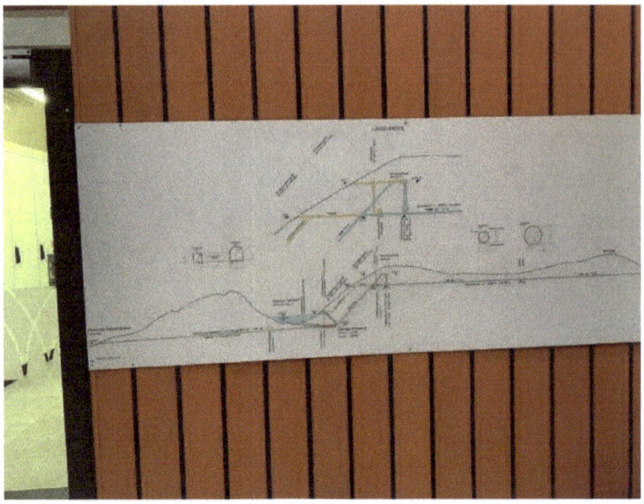

The generators are shown below. Each generator has a terminal nominal voltage of 25 kV DC (courtesy client).

Figure 8.2.2 Author pointing at the DC generator windings and cooling fans. (courtesy client)

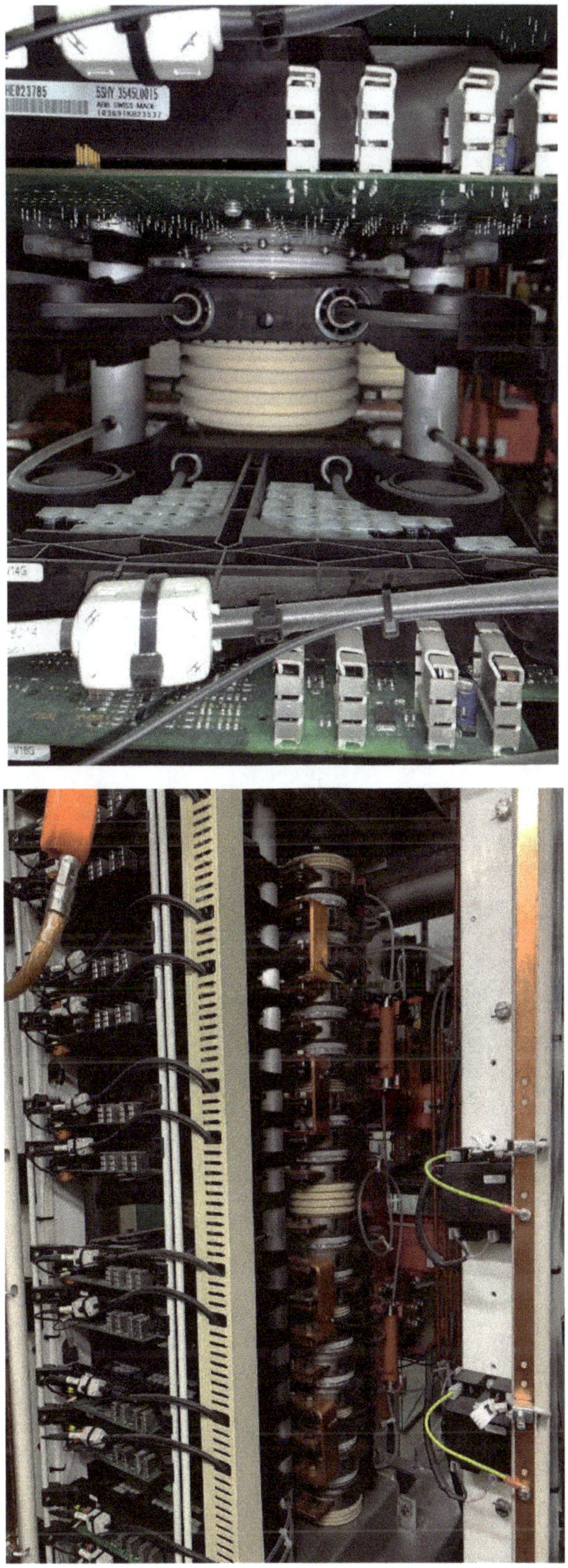

Figure 8.2.3 SCRs in the Inverter circuit – DC to ac (courtesy client)

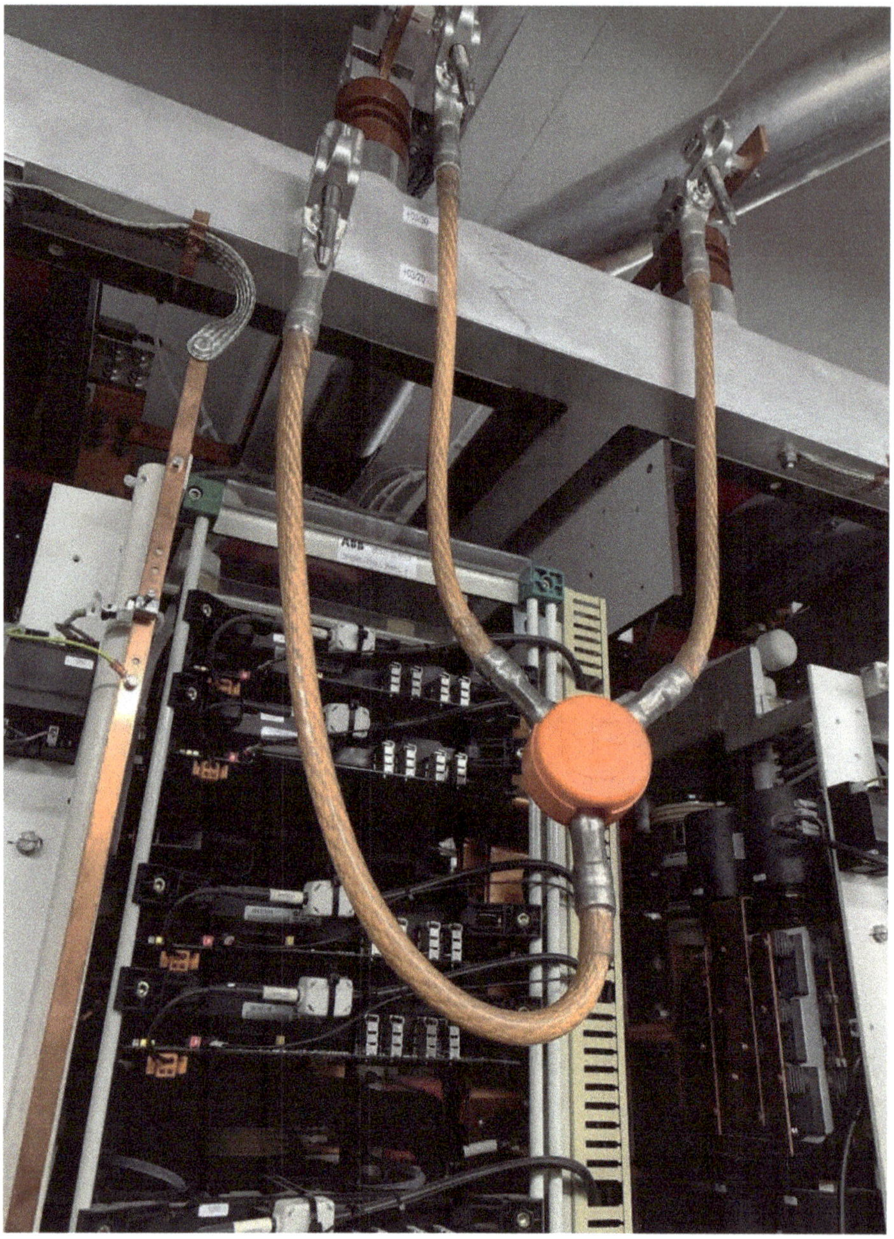

The three lines shown in the Figure above, as seen in reality in a Y format configuration. (courtesy client)

Figure 8.2.4 Inverter circuit (courtesy client)

The higher-level AC voltage is then collected by four AC Transformers, as shown in Figure below:

Figure 8.2.5 Core-Coil Assembly of the four transformers (courtesy client)

These transformers were tested by our patented FDD-EPT PD detection and location system by Diagnostic Devices Inc. (US PTO 6,178,386). Some live-PD signals are shown in the Figure below. The signals are obtained by ultrasonic piezoelectric sensors, also patented by Diagnostic Devices Inc. These sensors need a 28-volt DC power supply, which is obtained from resident 5 volts DC source using a boost DC-DC converter described below in the circuit diagram.

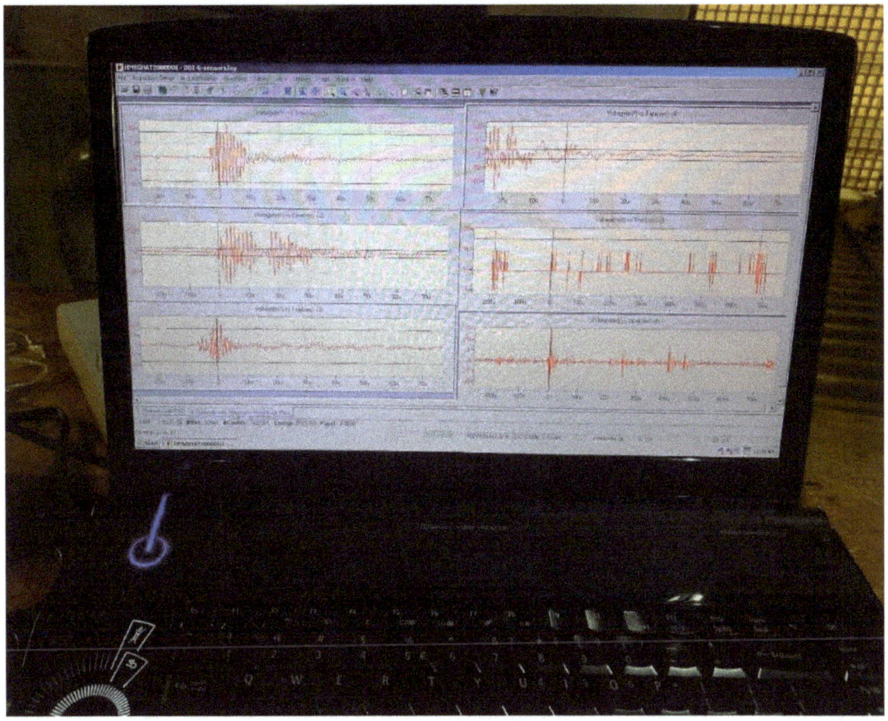

Figure 8.2.6 PD Signals measured using FDD-EPT for transformer at Grimsel-II (courtesy DDI)

The measured signals clearly show PD signals on four of the attached sensors on the surface of the transformer. These signals will then be massaged by the IIMSGHAT 2000 patented software in conjunction with FDD-EPT to generate a 3-D diagram of the transformer tank and the estimates of the failure locations within the confines of the transformer tank with a minimum guarantee and confidence level of 85% of the failure caused by PD.

Figure 8.2.7 PD Signals measured using FDD-EPT sensors for transformer at Grimsel-II (courtesy DDI)

Chapter Problems on Application of Boost DC-DC Converters to Acoustic Sensors of FDD-EPT

Problem P.8.1:

The supersonic piezo acoustic sensor that collects PD signals needs a 28-volt DC power supply. The resident power source of FDD-EPT system has a maximum DC volt power supply of 12 volts. Design a suitable single-ended amplifier-filter-for the DC-DC Converter system using Boost converter as discussed earlier and as in Chapter 3 to be suitable for 12 volts DC to 28 volts for the sensor, an amplification factor of 200 and a ripple voltage of less than 2%. Use suitable MOSFETs, diodes, and BJTs to achieve this new system for the above purpose.

Problem P.8.2:

In Problem 1 above, use PSpice or MuitiSim to build the hardware system. System should be suitable for 12 volts DC to 28 volts for the acoustic sensor, with an amplification factor of 200 and a ripple voltage of less than 1%. Use suitable MOSFETs, diodes and BJTs to achieve this new system for the above purpose.

Case Study III - Interleaved Input Series Output Ultra-High Voltage Boost DC-DC Converter; 12v Dc Input and 160-220 Volts Output

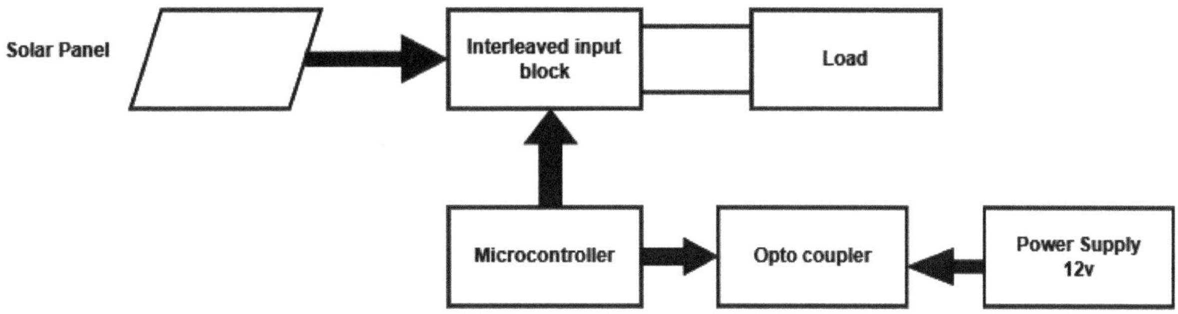

Figure 8.2.2

The input to the circuit is a solar panel of 12 volts DC and 15-watt capacity. The solar input is fed to the Interleaved input block. The circuit also consists of a microcontroller, an opto coupler block, and the power supply to the opto coupler. The microcontroller block uses the buffer to amplify the current gain.

The operating voltage of the Microcontroller is 5 volts and is same for the buffer is the same. The optocoupler is an MC 2TV with a power supply of 12 volts.

A 12-volt input generates a 160-220 volts ultra-high boost output.

CASE STUDY IV Turn-on and Turn-off Snubber circuits for switch protection in Power Electronic DC-DC Converter Circuits

The components of DC-DC Converter circuits that employ diode, SCRs, ICBTs, or ICET for switching actions for buck-boost, fly back, Cuk isolated or un-isolated DC-DC Converter circuits encounter very small power loss levels when it is on or off, but when the switching takes place the power dissipated is not so benign. It could rise to unwanted levels that could lead to failure due to EMI. All these switching devices operate in the Safe Operating Area (SOA) when on or off. However, in the transition between switching on and off, there are several dv/dt or di/dt conditions due to change in operating conditions.

These lead to the turn-on and turn-off conditions that need protection and, which is provided by what are called the snubber circuits. These use passive components, capacitors, and inductors along with resistors to design circuits for such switch protection by keeping the energy dissipated during action as low as possible. These devices are called as 'Snubbers' and are of two categories:

1) Turn-off Snubber – controls rate of rise of voltage, related to the dv/dt – associated with the use of a capacitors
2) Turn-On Snubber – controls rate of rise of current, related to di/dt – associated with the use of an inductor

These actions are shown in the Figure below for a turn-off snubber:

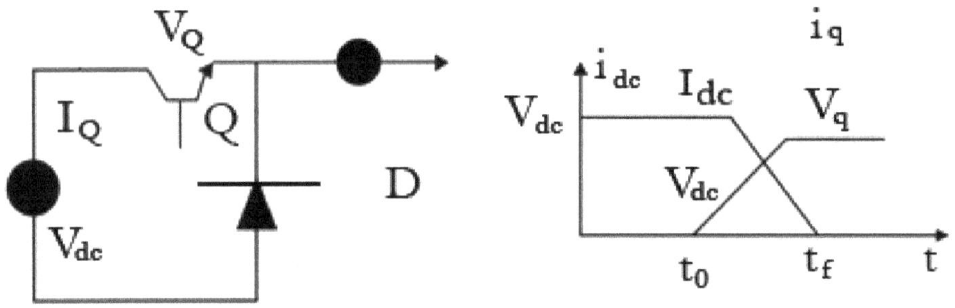

Figure 8.4.1 Transistor as a switch complemented by a diode.

We can now calculate the Turn-off dissipation and the related SOA as follows:

This is given by the product of the V_{dc} and I_{dc} and is given by the following equation:

$$W_{dissp} = \int_0^{t_f} v_Q\, i_Q dt = \frac{V_{dc} I_{dc} t_f}{2} \qquad (8.4.1)$$

So, say for a switching device

$I_{dc} = 50$ A

V_{dc}=400 V

t_f=0,5 μs

Then the energy lost is about 5 mili Joules. If the switching frequency is 20kHz, 100 Watts would be dissipated in the transistor of the turn-off transition alone. If we superimpose this dissipated energy on the SOA of the transistor, as in the Figure 8.4.3 below,

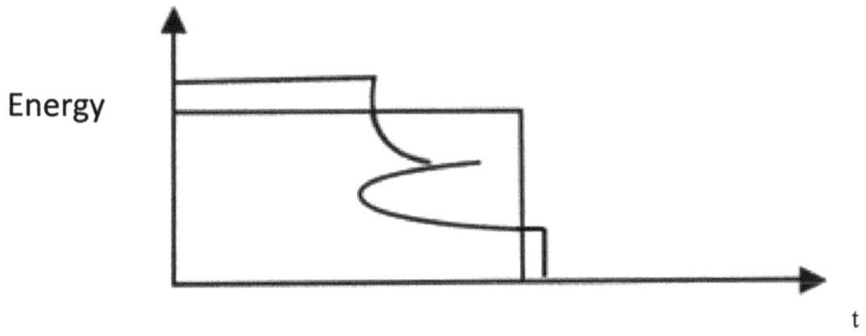

Figure 8.4.3 Energy of a SOA

Figure 4 SOA of a turn-off snubber circuit it will be seen that the safe operation is not possible. In order for this to happen, a suitable circuit using a capacitor is used as shown below:

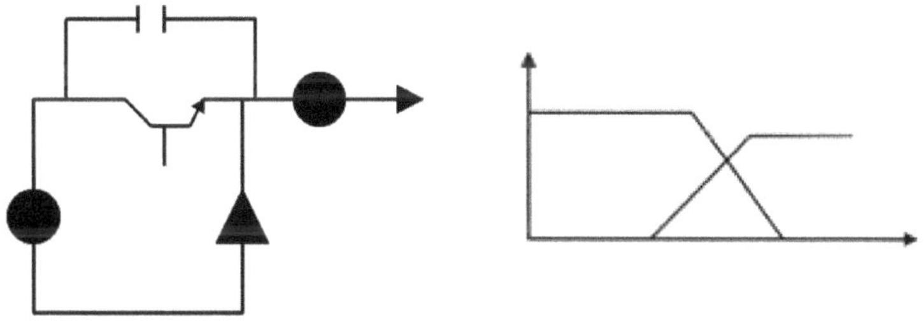

Figure 8.4.4 SOA of turn-off circuit with a capacitor

With the introduction of a capacitor across the BJT creates a new SOA, fresh turn-off circuit locus and turn-on circuit locus, the turn-off circuit current, and voltage across the BHT switch as in the figure below.

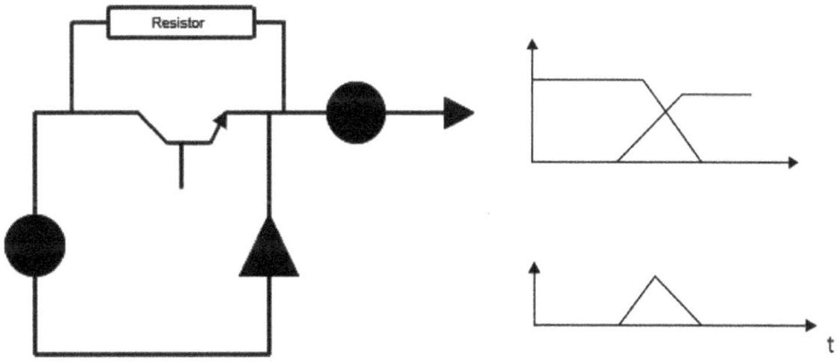

Figure 8.4.5 SOA with circuit operation in the safe mode of operation

This can be explained with an example as follows:

Example

Figure shows a capacitor C_s across the BJT, transistor voltage and current waveforms at turn-off. Power dissipated in the transistor circuit as a function of time during turn-off. All aspects are shown in the +figure above.

Choosing a value for C_s with new t_f'

Value of power now dissipated is given by:

$$W_{dissp} = \int_0^{t_f'} v_Q i_Q \, dt = \int_0^{t_f'} \gamma_{t_f'} V_{dc} (\frac{1}{t_f'})^2 I_{dc} \left(1 - \frac{1}{t_f'}\right) dt \quad (8.4.2)$$

$$= \frac{1}{12} \gamma_{t_f'} V_{dc} I_{dc} t_f' \quad (8.4.3)$$

Choosing:

$I_{dc} = 50$ A, $V_{dc} = 400$ V, $\gamma_{t_f'}$=0.2, $t_f' = 0.33\ \mu s, and\ W_{dissp} = 0.12\ mJ$;

During the time $0 < t < t_f'$, the capacitor current is given by

$$i_C = I_{dc} - i_Q = \frac{I_{dc}}{t_f'} t \quad (8.4.4)$$

The charge in the capacitor Q_C delivered to C_s in the time t_f' is the integral of this current between 0 and t_f'. We can now calculate the value of C_s by solving the following equation:

$$C_S = \frac{Q_C t_f'}{\gamma_{t_f'} v_Q} \quad (8.4.5)$$

$$C_S = \frac{I_{dc} t_f'}{2\gamma_{t_f'} V_{dc}} = 0.1 \mu F \quad (8.4.6)$$

A mere capacitor in parallel with the BJT does not serve practically to save the switch due to the discharge of the charge through the BJT which may have low inherent capacitance and may damage it.

To avoid this, a resistance and diode are used in a parallel combination in series with this capacitor C_S.

The value of R_s can be suitably found to reduce the dissipated power during switching.

The Turn-on Snubber

When the converter tries to change its state to the 'on' condition, the converter current and voltage are shown in the figure below. Here, the voltage V_{dc} drops to zero, and the current increases to its maximum DC value and stays there i_Q,

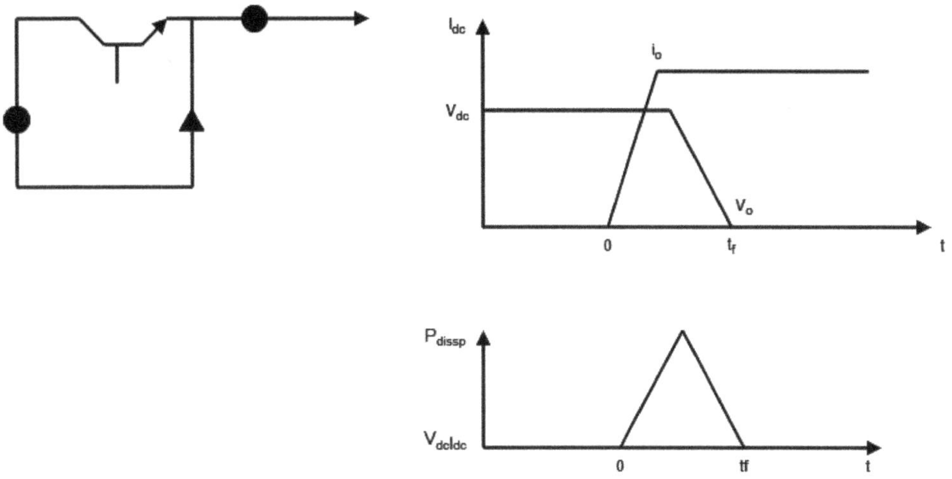

Figure 8.4.6 Turn-On Snubber circuit with converter transition voltage and current

In the turn-on condition, the power dissipated in the switch is given by:

$$W_{dissp} = \left(\frac{1}{12}\right)\gamma_t V_{dc} I_{dc} t_f' = 0.06mJ \qquad (8.4.7)$$

These are all inductive currents and are proportional to $1/L_s$. We can choose appropriate values of L_s to limit the transistor current to a fraction γ_s of L_s by the time v_Q reaches zero. Assuming values of γ_t =0.33 and t_f' =0.1 micro sec, yields the value of dissipated power as above. We can protect and lower the power dissipated by using an inductor in series with the BJT as shown below.

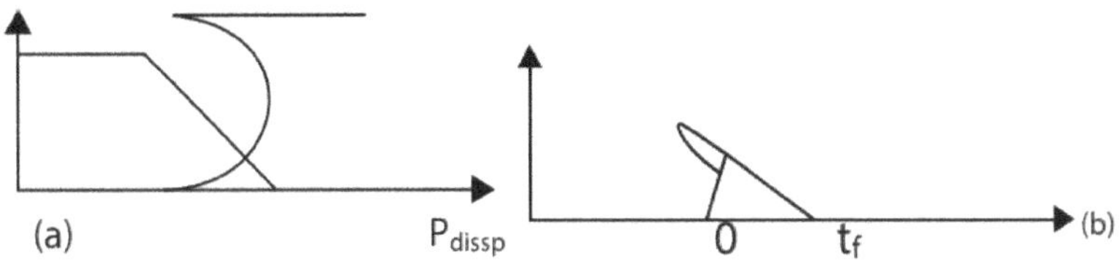

Figure 8.4.7 Turn-on Snubber with an inductor to limit power dissipated during transition time.

This is further tuned by providing a series combination of resistance and diode put in parallel with the inductor that is in series with the BJT. This can be explained with an example that illustrates the choice of the component values of the final assembly of the turn-on snubber circuit to limit the operation within the SOA and limit the power dissipated to a low value to protect the BJT. Now, at a switching frequency of 20 kHz the resulting power dissipated in the BJT at switching on is about 1.2 W. This a almost two orders lower than the power dissipated without this inductive snubber. The peak power is also reduced, and the turn-on switching locus now remains within the

SOA, as shown in the figure below (c) above. A practical turn-on snubber contains a combination of a resistance and diode in series across the inductor L_s as indicated earlier.

All this can be illustrated with an example where, finding the element values for the turn-on snubber circuit can be found. We will try to find the value of L_s from the relationship of flux linkage as follows:

$$\lambda = L_s i \quad (8.4.8)$$

At $t = t_f'$, we will have

$$i_Q = \gamma_i I_{dc} \qquad (8.4.9)$$

The flux linkage value can be found by integrating the voltage across the inductor v_L

$$\lambda = \int_0^{t_f'} v_L dt = \frac{V_{dc} t_f'}{2} = L_s \gamma_i I_{dc} \qquad (8.4.9)$$

Which yields

$$L_s = \frac{V_{dc} t_f'}{2\gamma_i I_{dc}} = 1.2\, micro\, H \qquad (8.4.10)$$

The final value of L_s of 1.2 micro Henry is based on our previous parameter values of $V_{dc} = 400$ V, $I_{dc} = 50$ A, $Y_i = 0.33$ and $t_f' = 0.1$ micro sec. If we expect the shortest time that we expect the transistor to be off is 2.5 microseconds or 5% of a 20k Hz cycle. We need the time constant L_s/Rs of about 0.5 micro seconds to ensure that L_s is completely discharged. This needs a resistance value of $R_s = 2.5$ ohms.

Combined turn-off, turn-on snubber circuit

For both turn-off snubber and turn-on snubber circuits to operate simultaneously for a high voltage DC power like a DC-DC Converter, one has to employ both the snubber circuits explained above to keep the power dissipated in the power switch at the minimum so that the operation of the switch remains in the specified SOA. But, in practice, this is not necessary to be implemented, one can use the single combination of a resistance and diode to be in parallel or in series with the capacitor and the inductor elements to effect the same snubber circuit result as both the combinational circuits. This can be explained as follows by looking at the operation of the circuit in figure below:

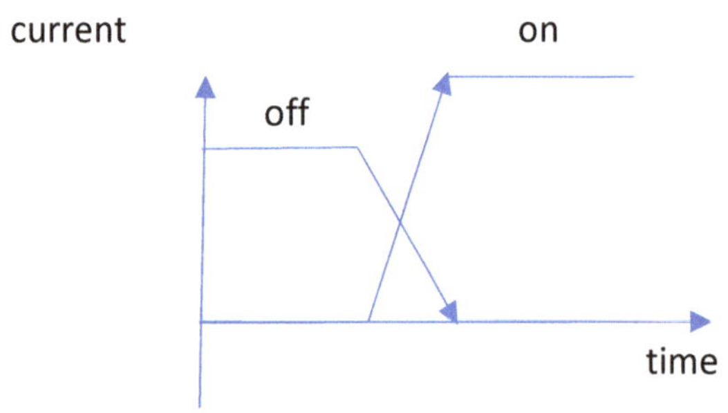

Figure 8.4.8 Practical turn-off/turn-on snubber circuit

The circuit components C_s, R_s, D_s (especially the latter two) L_s operate for a turn-on snubber circuit and in its parallel combination as a snubber for the turn-off snubber circuit operation combination as a part of the turn-off snubber circuit. The only difference is in the time-dependent operation of the two circuits. From time 0 to t_1 and later from t_1 to t_f'. The operation of the circuit in the turn-on snubber operation but the discharge current of C_s should be added to i_Q. However, in the turn-off circuit, the operation is similar to the previous circuit between time-period 0 to t_1, after which the capacitor C_s is discharged to V_{dc}. But at t_1 when $v_C = V_{dc}$ and D turns on, L_s is still carrying a current $i_L = I_{dc}$ and L_s and C_s will resonate for quarter of a cycle time until $i_L = 0$. At this time, D_s turns off, and R_s damps the transient. The circuit waveforms are shown in the Figures below. The incremental ringing voltage across switch Q is given by:

$$\Delta v_A = I_{dc}\sqrt{\frac{L_S}{C_S}} \qquad (8.4.11)$$

This incremental voltage is added to V_{dc} to give the peak transient voltage, which is the peak voltage across C_s. This can be illustrated with an example.

In the previous examples of turn-off snubber and turn-on snubber circuits, we have calculated values of C_s =0.066 micro Farads and L_s = 1.2 micro Henrys respectively. These values will help us to calculate the peak BJT voltage V_Q as follows:

$$V_Q = V_{dc} + \Delta v_Q = V_{dc} + I_{dc}\sqrt{\frac{L_S}{C_S}} = 400 + 50\sqrt{18} = 613\ volts \qquad (8.4.12)$$

These are extremely large values of V_Q relative to V_{dc}. Thus, the choice of the passive components have to be done judiciously to limit these voltage values.

Another way of implementing the turn-on/turn-off snubber is shown in the figure below, where a resistance in series with a diode (R_s, D_s) is placed in parallel across the inductor L_s, and a capacitor C_s is placed across the entire series combination of BJT and this snubber circuit of the turn-on snubber. The switch voltage v_Q and switch current i_Q are identical as before, except the discharge voltage across R_s contributes to v_Q. The peak current can now be calculated in two stages as follows from 0 to t_1 and t_1 to t_2 as follows:

$$I_Q = I_{dc} + \Delta i_Q = I_{dc} + V_{dc}\sqrt{\frac{C_S}{L_S}} \qquad (8.4.13)$$

This is so because at turn-on the circuit exhibits a period of second order behavior. After $v_Q = 0$ at t_f', L_s charges linearly to I_{dc} at t_1, at which time D turns on. During this period a D is reverse biased and $v_C = V_{dc}$.

After D turns off at t_1, L_s and C_s resonate until $v_C = 0$, and D_s turns on $t = t_2$. Also, R_s comes into circuit i_Q settles to I_{dc} in an overdamped condition, as shown in the figure 10 below:

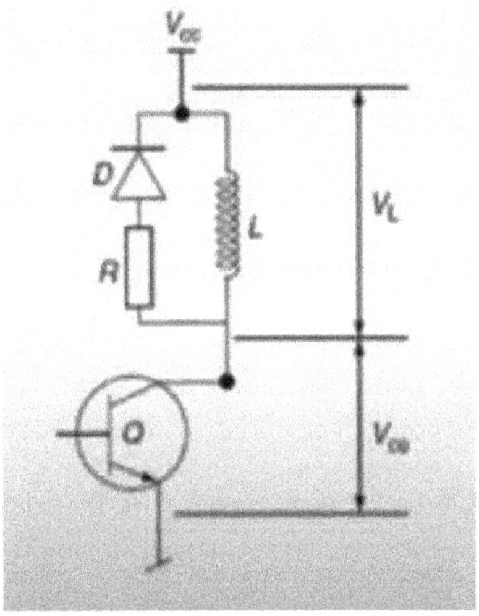

Figure 8.4.9 Alternate circuit for turn-on/turn-off snubber

Figure 11 shows a turn-off snubber circuit with the capacitor in series with a parallel combination of a resistance and diode; and combination thereof placed across the main diode. In a similar manner Figure (b) shows a turn-on snubber circuit with a series combination of a resistance and diode placed across the inductor, which in turn is placed in series with the main diode.

Power dissipated in Snubber circuit:

1) Turn-off snubber circuit – energy stored in the snubber capacitor is given by:

$$W_C = \frac{1}{2} C_s {V_{dc}}^2 \qquad (8.4.14);$$

which comes out to be for a suitable Y = 8.25 mJ for the previous example discussed.

2) Turn-on snubber circuit – energy stored in the snubber inductor is given by:

$$W_L = \frac{1}{2} L_s {I_{dc}}^2 \qquad (8.4.15)$$

which comes out to be for a suitable Y; = 1.5 mJ for the previous example discussed.

3) Turn-of, turn-on snubber circuit – energy stored in the snubber capacitor and inductor is given by:

$$W_C + W_L = \frac{1}{2}\left(C_s {V_{dc}}^2 + L_s {I_{dc}}^2\right) = 8.25\ mJ + 1.5\ mJ = 9.75\ mJ \qquad (8.4.16)$$

Increased power dissipation in snubber circuits

Both the turn-off snubber circuit and the turn-on snubber circuit are considered a single-switching element the corresponding turn-off or turn-on voltage and current waveforms for these respective elements indicated are shown for the turn-off snubber in the figure below:

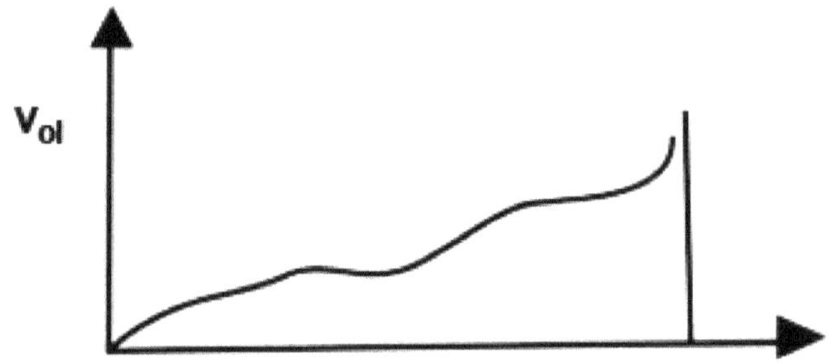

Figure 8.4.10 Waveforms for turn-off snubber

The related power in the entire circuit is given by the Figure ©. Both the switch and the diode share the voltage and current rise, so initially, the voltage of the switch rises, and when it reaches V_{dc} the diode turns on and picks up the load current I_{dc} the total energy absorbed is V_{dc} I_{dc}. This energy is t_f' $(2t_f'Y)$ larger than without the snubber. Similar increase in energy absorbed by the tun-on snubber can be seen resulting in larger energy dissipated in the entire snubber element.

Snubber circuits that can recover energy lost in the switch element

Using the previous equations and with a switching frequency of 20 kHZ, the total power dissipated by a combined turn-on/turn-off snubber circuit is given by

$$P_{dissp} = 20x10^3[\frac{1}{2}(C_sV_{dc}{}^2 + L_sI_{dc}{}^2) = 20(8.25 + 1.5) = 195\ watts \quad (8.4.17)$$

This is 3.9% of the total power output of the converter, with a 25% duty ratio. An alternate circuit to recover this energy is shown in Figure 12 below:

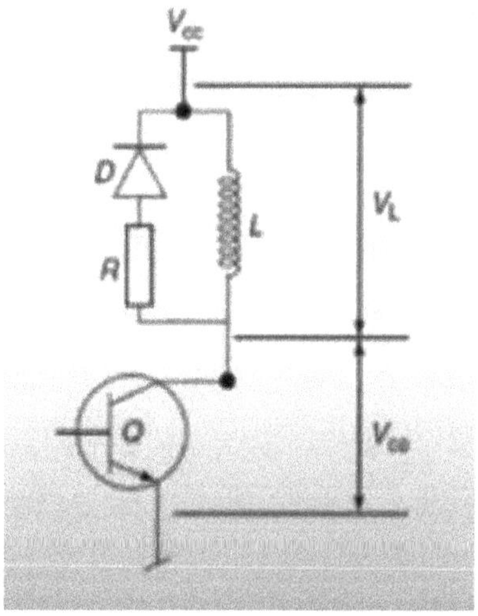

Figure 8.4.11 Energy recovery turn-off snubber circuit diagram

The operation of this circuit can be explained in detail and is discussed here as follows for a turn-off snubber as in Figure above.

When the switch Q is on the two capacitors C_1 and C_2 charge to a voltage of V_{dc}. When Q is off, the two capacitors, along with their diodes, are now like two turn-off snubbers in parallel with the inductor, and they are not in the circuit at all because the diode D_3 I is reverse biased. The net snubber capacitance is the sum of the two capacitors $C_1 + C_2$.

When the switch Q is not turned on, its current rises to I_{dc} in time t_r', during which time D turns off and v_Q falls to zero in time t_r''. If t_r'' is small compared to the period of ringing in the snubber circuit, then the series circuit consisting of f C_2, D_3, L, and C_1 is excited by a voltage step V_{dc}. Now D_3 turns on, and the response of the series circuit is a second-order ringing condition. The ringing continues for one-half of a cycle, at which time i_L tries to go negative, turning off D_3. At this point, the total capacitor voltage rises to $2V_{dc}$, and the individual voltage across each capacitor depends on the values of C_1 and C_2. For $C_1 = C_2$, this individual voltage is equal to V_{dc}, and the circuit is set for the next turn-on cycle. It is imperative to note at this point that the combined snubber circuit has not completed a full cycle of its operation and does not need to dissipate its energy.

In addition, one can also have lossless snubber circuits, which are beyond the scope of this book.

Chapter Problems on Snubbers

P.9.1 Noise reduction in HVDC-DC Converters

It is imperative to know why does one need low noise HV DC-DC converters.

1) Transducer Signal Conditioning
a. Photomultipliers
b. Avalanche Photodiode operation
c. Ultrasonic applications
d. Use of sensitive Microphones
e. Radiation detectors

In such cases, generally for low noise typically, noise levels are of levels:

< 1 mili Volts, noise/100MHz BW @ 100-1000 volts DC output with a load of < 5 mili amps and as low a micro amp with a low voltage supply.

A. An output stage runs through two-section filter . Draw the Figure, with modulation through a step-up transformer and feedback current control through a resonant sine wave converter:

In this, two-section filter, it is essential to notice that a resistor is used in the filter network instead of an inductor. This is because a small voltage drop across a resistor in the case of large DC volts out is tolerable. In addition, the resistance has a low shunt capacitance that is useful in such a much better filter. One problem with such a low noise controller is the expense of power in the control circuit element, which can be bypassed by using a low power switching controller like LT 1534, which is an edge time switching control regulator that maintains low harmonic content in switch mode delivery drive as shown in the circuit above.

B. Push-Pull pulse width modulator

In the circuit above, we can get rid of the entire bottom end with edge time control with step-up transformer modulation through the same resistively coupled resistor filter circuit, down to the feedback network to complete the network. This can lead to a better performance of the controller for low noise control for an HV DC-DC converter.

Such circuits can yield as low as 100 microvolts peak to peak noise level for a 1000-volt DC output converter with a five-volt input supply. This is pretty low level of noise for such high voltage DC-DC converters.

6 Types of Chopper Circuits

Types of Chopper Circuits

Chopper circuits and unidirectional power semiconductors are frequently used. If the choppers are properly arranged in the circuit, then all four kinds of choppers as mentioned below are possible, with operation in all four quadrants of its i-v characteristics. There are four types of chopper circuits:

- Class A – 1st quadrant
- Class B – 2nd quadrant
- Class C – 2 quadrant operations
- Class D – 2 quadrant operations
- Class E – 4 quadrant operations

Class A has i-v characteristics are in the first quadrant so suitable for DC motoring action controls.

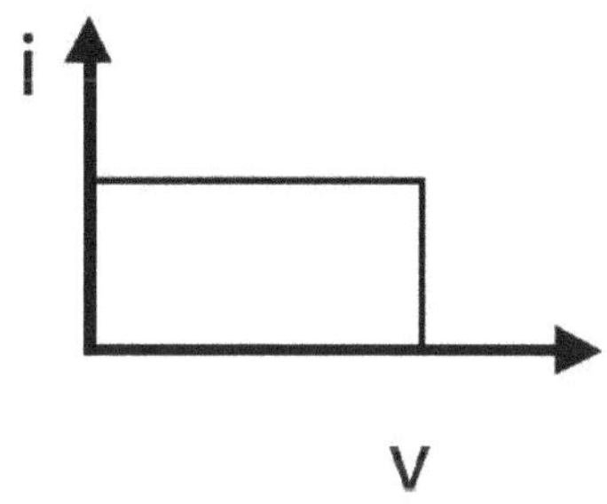

Figure 8.5.1 1st quadrant operation – Class A

Class B i-v characteristics are also one quadrant operation but related to breaking action control of DC motors.

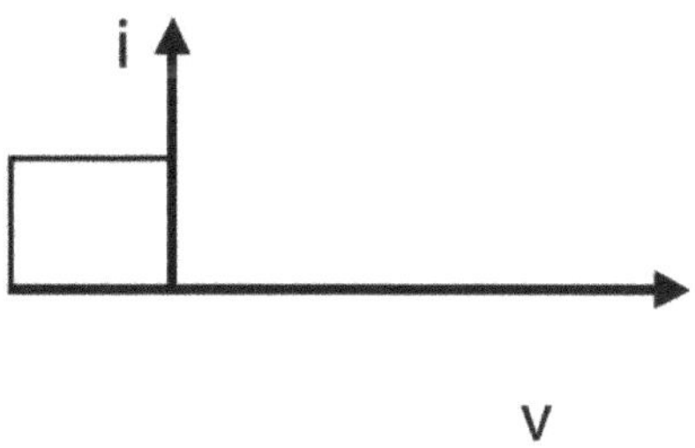

Figure 8.5.2 2nd quadrant – Class B

Class C and D are related to the forward and reverse direction of operation, respectively, while Class E is associated with both directions of operation for motoring actions controls. Class C with regenerative action will become Class E.

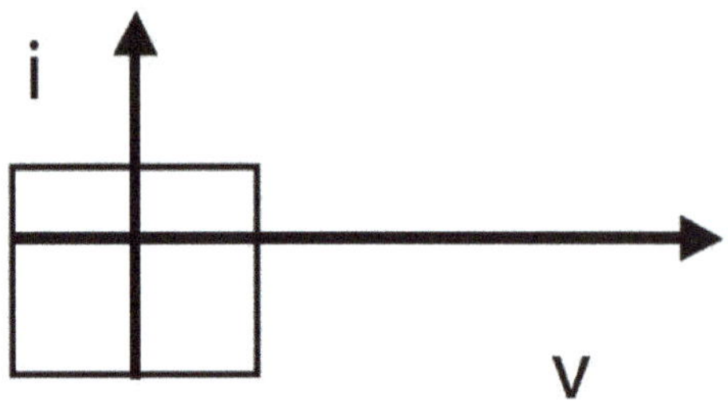

Figure 8.5.3 i-v characteristics

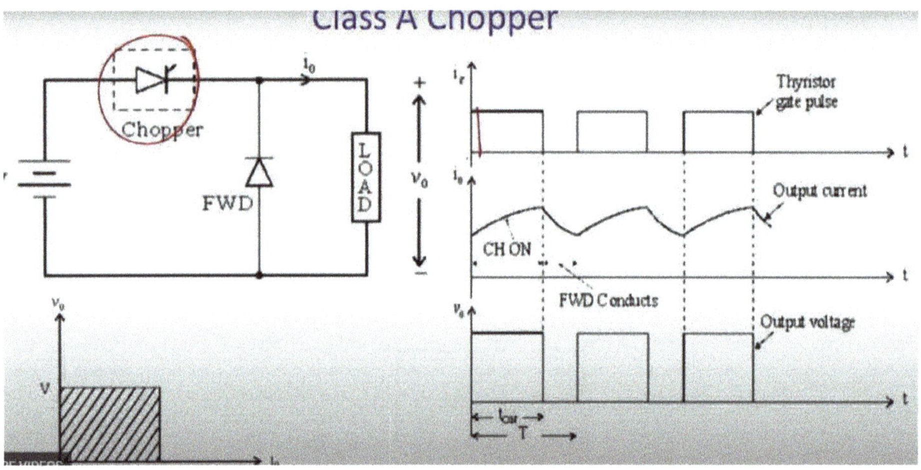

Class A chopper

Figure 8.5.4 Class A chopper circuit or 1st quadrant chopper (chopper is highlighted)

A class A copper works as follows:

1) When chopper is on, supply voltage is connected across load, which yields

$$V_0 = V_z$$

2) When chopper is off,

$$V_s = 0$$

And load current continues to flow in the same direction due to inductive load through the freewheeling diode.

3) The average value of voltage and current are always positive, so Class A chopper is called 1st quadrant chopper as well.
4) It is also called a step-down chopper as power flows from source to load.
5) This kind of chopper is used to control speed of a motor.

Class B Choppers

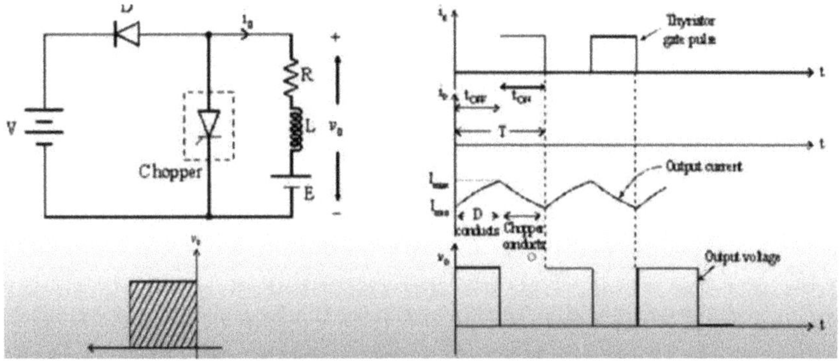

Figure8.5. 5 Class B Chopper circuit

1) When chopper is ON, the DC source (battery) E drives a current through the inductor L and resistor R in an opposite direction.
2) During ON period of chopper, the inductor L stores energy. So, output voltage

$$V_0 = 0$$

3) When chopper is OFF, diode D conducts, and a part of the energy stored in inductor L is returned to the supply. So output voltage is given by:

$$V_0 = E + L(di/dt)$$

4) Thus, average output voltage is positive but average output current is negative.
5) Therefore, class B chopper operates in the 2nd quadrant
6) In this chopper, power flows from load to source and thus is also called the step-up chopper.

Such choppers are used for regenerative breaking of DC motor.

Class C Choppers

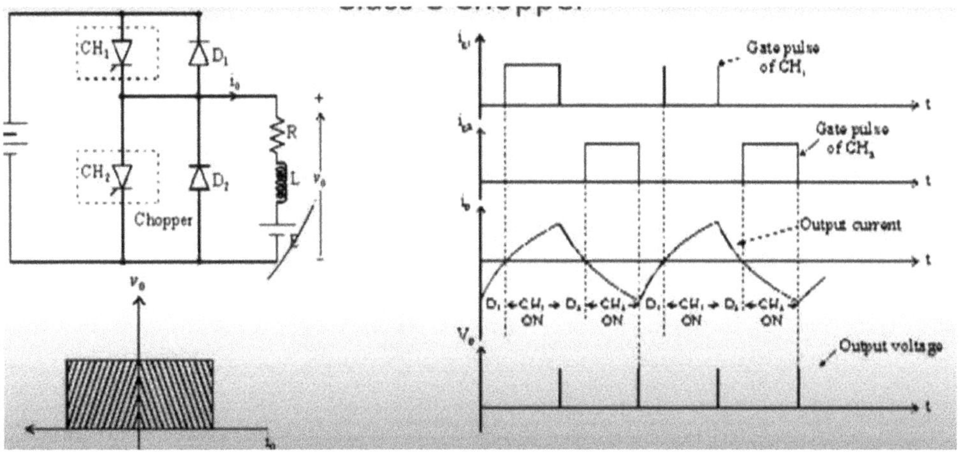

Figure 8.5.6 Class C chopper circuit and performance (see red line indicator)

1) Lass C chopper can be used both as step-up or step-down chopper.
2) Class C chopper is combination of Class A and class B choppers.
3) For first quadrant operation $CH_1$is ON or D_2 conducts; for 2nd quadrant operation CH_2 is ON or D_1 conducts.
4) When CH_1 is ON, load current is positive. In ON time interval output voltage is V and power flows from load to source.
5) When $CH_1$is turned OFF, energy stored in inductor L forces current to flow through the diode D_2. So, output voltage becomes zero. Then, current continues to flow in positive direction.
6) When $CH_2$is triggered, the voltage E forces current to flow in opposite direction through L and CH_2. Therefore, output voltage is 0.

Class D chopper

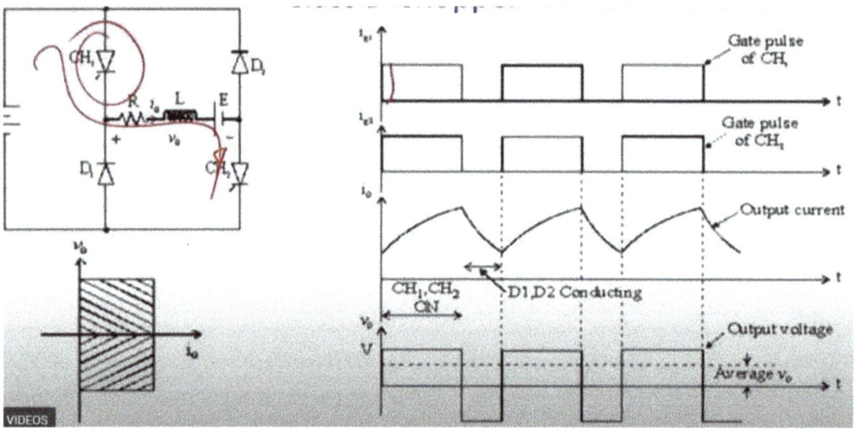

Figure 8.5.7 Class D chopper circuit and working. (see redline passage of current)

1) Class D chopper is also two quadrant choppers.
2) When CH_1 and CH_2 are ON at a time, output voltage V_0= V, and output current flows through the load.
3) When CH_1 and $CH_2$are turned OFF, load current continues to flow in the same direction through load, D_1, D_2 due to inductive load. SO output voltage V_0-= - V
4) Average load voltage is positive if chopper ON time is more than OFF time and vice versa.
5) Therefore, load current flows always in positive direction, but load voltage can be positive or negative.
6) This type of chopper operates in both 1st and 4th quadrants.

Class E Choppers – four quadrant operations

Problems on Choppers

Basics of Switching Power Supplies

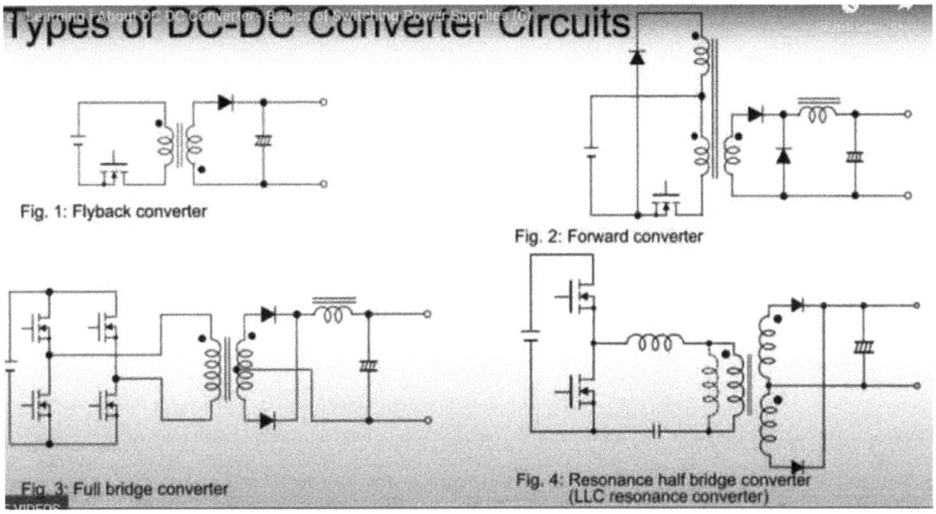

Figure 8.5.8 AC-DC Converter circuit with full-bridge rectifier

AC-DC Converter has four basic blocks:

1) Input Filter Block – A switching element of switching supply produces high frequency noise. This input filter block reduces this noise.
2) Bridge Rectifier – The AC is rectified by the bridge rectifier to provide almost DC voltage at its output.
3) DC-DC Converter block -There are many topologies for the DC-DC converter block depending on the requirements of the application. These could be buck-boost, flyback, SEPIC, etc.
4) The feedback control block – This block controls the output voltage and controls the switching time of the MOSFET on the primary side.

We will now discuss the different DC-DC converter circuits

a) Fly back converter
b) Forward converter
c) Full bridge converter
d) Resonance half-bridge converter (LLC resonance converter)

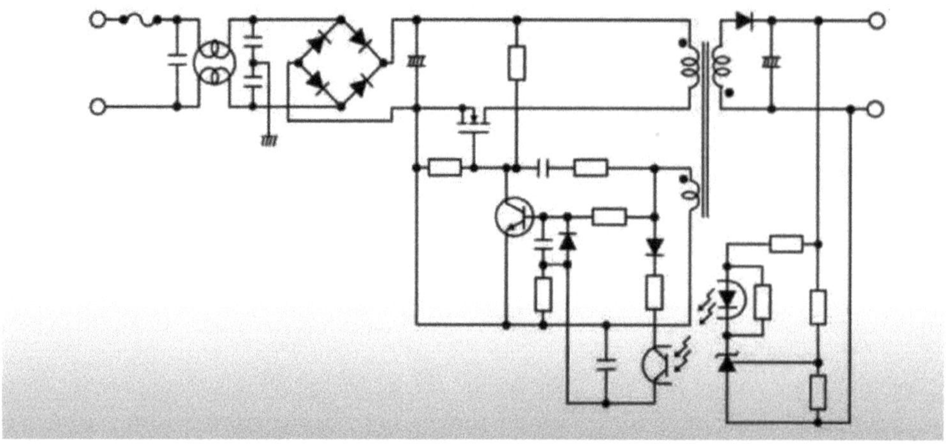

Figure 8.5.9 DC-DC Converter circuits

We will explain a fly back converter, and forward converter will be explained here. A fly back converter is used up to 120 watts, a forward converter is used up to 500 Watts, a resonance half-bridge converter is used up to 1.6 kW, and a full bridge converter is used up to 1 kW or more. The following table lists the characteristics of each circuit topology.

Table 1. Characteristics of different DC-DC Converter circuits

	Type of circuit	Power level	Advantages	Disadvantages
1	Flyback Converter	Up to 120W	The circuit consists of the minimum number of parts	Large amount of electricity is used. Efficiency decreases. A transformer is large.
2	Forward Converter	100-500 W	Superior to a Fly back circuit in efficiency	A transformer reset circuit is required.
3	Resonance half bridge converter (LLC resonance converter)	100 W to 1.6 k Watts	High Efficiency Low noise	A custom-designed transformer is required. Control of voltage is difficult.
4	Full-bridge converter	1 kW or more	High Efficiency Capacity can be increased	The number of parts increases. Control of voltage is difficult

Problems on DC-DC Converters

Impact of Reverse Power Flow on DC-DC Converters and Related Transformers

Problem P.8.5.1

In some applications like automobiles, which are hybrid vehicles it is necessary to have bidirectional power flow. In the buck mode for charging the battery power pack and in the boost mode for running the motors for suitable traction. The efficiency of such DC-DC converters can be optimized by designing the converter by placing some components of the converter circuitry in different locations. Conventional unidirectional power flow is where power comes from a source that is a generator and is converted to the transmission level through a step-up transformer and further converted to distribution level voltage through step-down transformers, then feeding the power to the load, as seen in Figure 1.

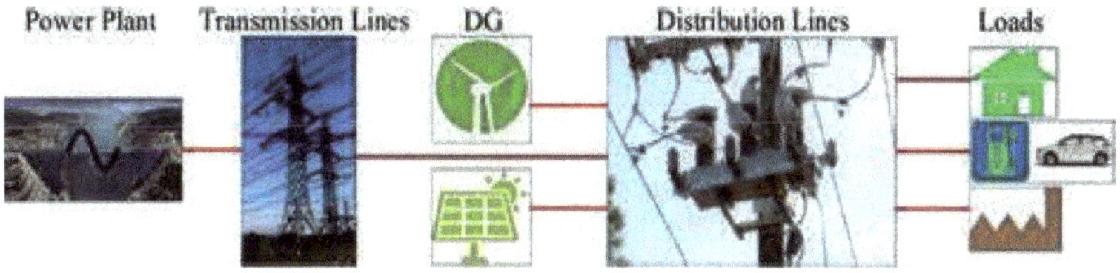

Figure P.1. Conventional power flow block diagram.

Problem 8.5.2

In a bidirectional power flow, power can be moved from source to load side (step-down or buck) and from the load side to the source side (step-up or boost), as seen in Figure 2.

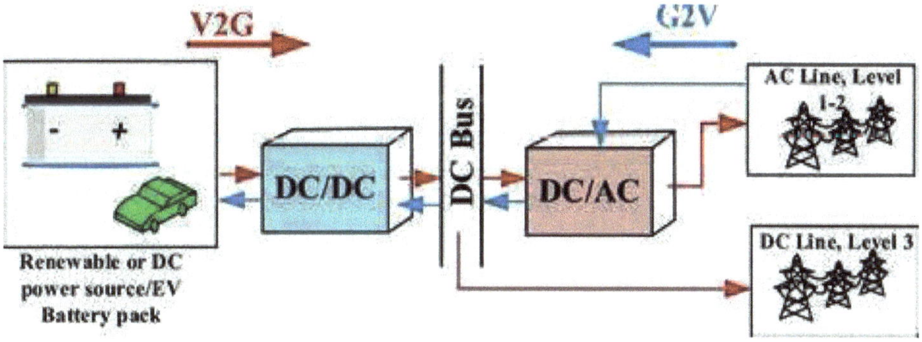

Figure P.2. Bidirectional power flow block diagram.

Here, the Distributed Energy Resources (DER) provide reverse power flow to the bidirectional power flow system. The DERs could be wind or solar, primarily if these are renewable energy sources. Before we go to understand the use of DC-DC converters in different kinds of DC voltages, we need to define these as follows:

1) High Voltage Direct Current (HVDC) 225 KV and above up to 1500 KV
2) Medium Voltage Direct Current (MVDC) 69 kV up to and less than 220 kV
3) Low Voltage Direct Current (LVDC) 115/220 Volts up to and less than 69 kV

4) Residential Voltage Direct Current (RVDC) suitable for residential application
5) Industrial Voltage Direct Current (IVDC) suitable for industrial applications
6) Commercial Voltage Direct Current (CVDC) suitable for commercial applications

Renewable power sources like Wind Power sources and Solar Power sources operate at between 34.5kV to 69 kV. Due to reverse power flow, the DC-DC Converters also experience some adverse effects, especially to the diodes, IGBTs, and IGCTs used as switches in the buck-boost or any other configuration at higher voltages as per the above classification. The reverse creates a condition of peak inverse voltage for switches like diodes, IGBTs, and IGCTs. Besides, in normal operation, the commutation of current can also affect the operation of IGBTs and IGCTs. Commutation causes oscillations in voltages which are generally mitigated using snubbers. These snubbers also share voltages. Also, the converters are provided with controlled IGBTs and IGCTs. The controllability provides more features than to just transmit power. Advantages of HVDC are as follows:

1) Security of Supply
2) Remote RES Integration
3) Grid Resiliency
4) Operational Flexibility
5) Public Acceptance
6) Lower Power Losses, this means more power
7) Controlled Power Flow, which leads to more grid stability and versatility
8) Smaller Footprint, which leads to more sustainability
9) Security of Supply which allows possibility of interconnecting grids and also enabling bigger markets in the power sector.
10) Remote RES Integration, that helps connections to large-scale hydroelectric grid and offshore wind systems.
11) Grid resiliency leading to special protection schemes, increased stability margins, and support during grid restoration.
12) Operational Flexibility which helps in variable RES integration, and better use of existing HVAC system.
13) Public acceptance which leads to smaller right of way (ROW), undergrounding, and easier permits.

Table 1. History of HVDC, some milestones

1954	The world's first commercial LCC link with Mercury valves
1970	Thyristor valves replace Mercury valves for LCC links
1997	The World's First VSC installation
2005	The World's First VSC installation from shore transmission to a platform
2007	First VSC Offshore Wind connection project
2010	The World's Frist 800 kV UHVDC LCC link
2014	Complete 1100 kV UHVDC system developed
2017	VSC HVDC highest capacity 3,000 MW 640 kV, 2000 km line
2019	VSC UHVDC 1500 kV, installation
2020	?

Section 8.6 Types of HVDC Connections

1) Bipole

2) Monopole

3) Symmetrical Monopole which are generally VSC only

4) Back to Back (BtB)

5) Rigid Bipole (mostly LCC)

6) Hybrid, which are not common

Examples of HVDC stations

Dorsey Converter Station – two Bipoles

Figure 8.6.1 Dorsey Converter Station with two Bipoles

Bipole Configuration – figure Shown as VSC but can be LCC

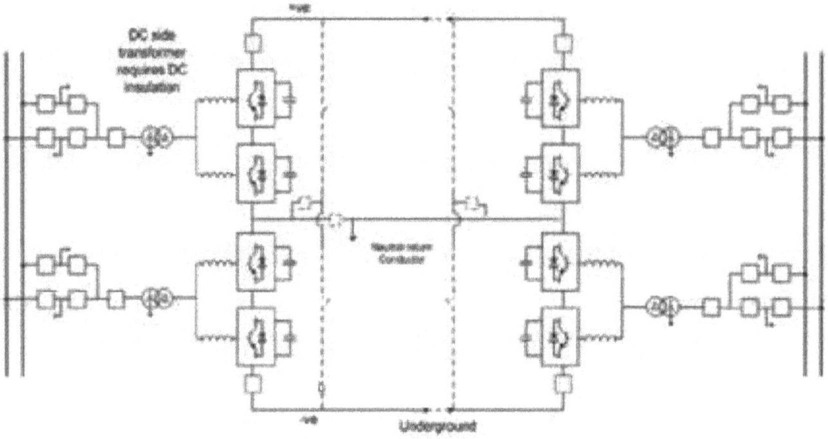

Figure 8.6.2 Bipole configuration as VSC but can also be LCC

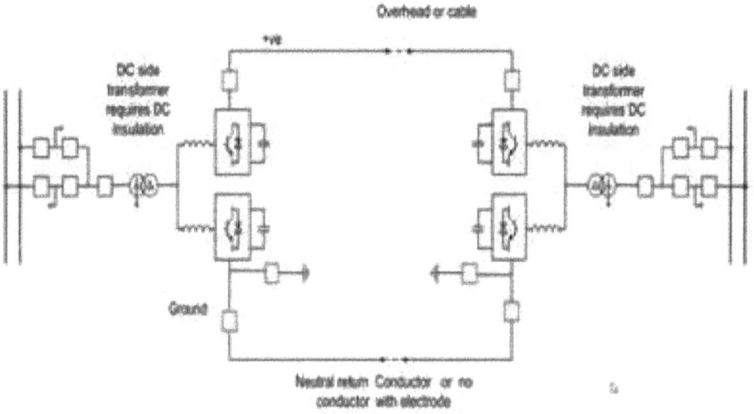

Figure 8.6.3 Monopole configuration as VSC but can also be LCC

Configuration Monopole

Generally, is also called an asymmetrical monopole and is operated with dedicated neutral. Such a connection is for low power levels, less than 1000MW, for undersea or BtB HVDC systems. The unwanted fact is the loss of full power during fault conditions.

An HVDC converter transformer is shown in the figure below:

Figure 8.6.4 A typical converter transformer for HVDC application at 2300 kV

A Symmetrical monopole is also shown below in Figure 6:

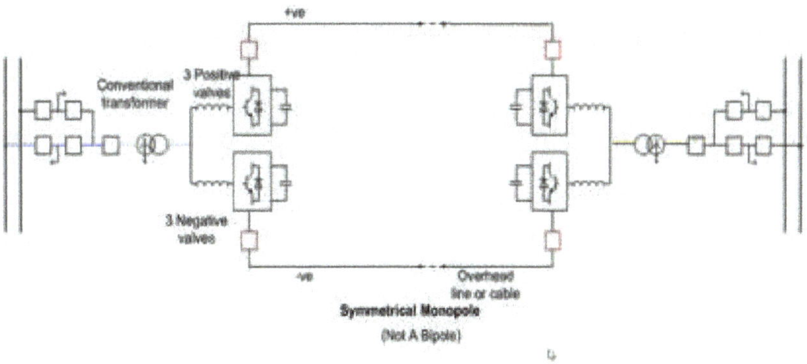

Figure 8.6.5 A typical Symmetrical Monopole configuration

In a monopole configuration half of the valves (diodes/IGBTs/IGCTs) are connected to the positive conductor and the other half to the negative conductors. There is no fault current for a fault to ground. If one conductor is grounded, the other conductor rises to a modulus of 2 pu voltage and must be cleared before the other flashes over. This is especially critical with the connected cables.

Back to Back (BtB) HVDC connection

There are many asynchronous systems between countries and even with such large countries like USA and Europe. Difficult to exchange between countries and fault levels could exceed ratings of Circuit breakers, DC-DC converters of all types. For such situations BtB HVDC connection is suitable.

Rigid Bipole HVDC System

A typical Bipole system is shown in Figure 7 below

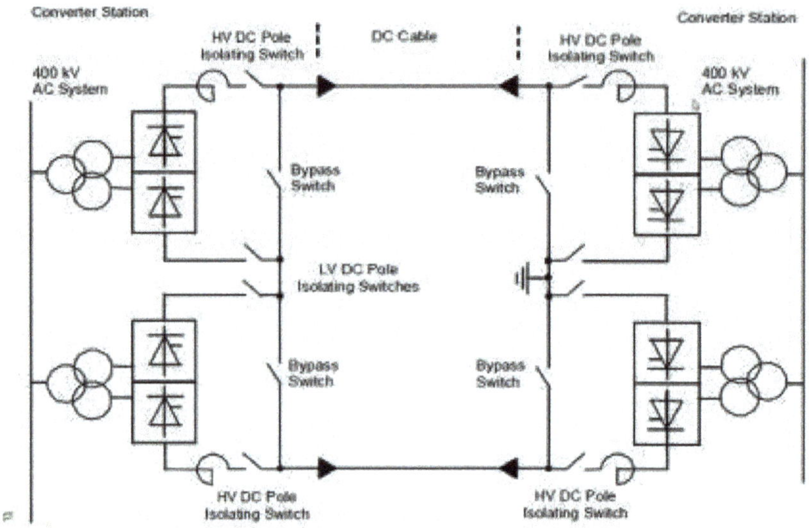

Figure 8.6.6 A typical Rigid Bipole configuration

Special HVDC voltage controller allows any drift in the current controllers to be compensated.

Proper insulation coordination and over-voltage minimization is also required. Only two HV conductors required saving costs if a cable is required.

HVDC Grids (MTDC)

HVDC Grids or MTDC are also generally called as multi-terminal HVDC as shown below in the Figure 8 below:

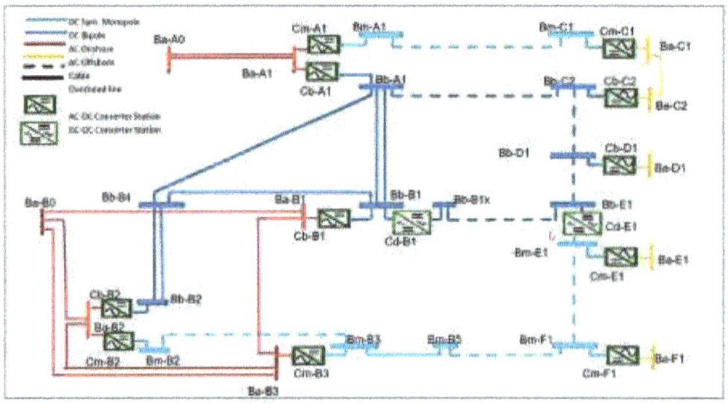

Figure 8.6.7 A typical HVDC Grid configuration (MTDC)

A lot of work is going on with five to seven multi-terminal HVDC systems. Besides, a feasibility considerable study has been conducted on a five terminal offshore system where the onshore transmission system is very congested. This shown in figure 8 above:

The fault clearing requires suitable CBs and DC-DC Converters or a full bridge topology rectifier for a VSC Configuration. Suitable HVDC CBs have been developed along with suitable DC-DC Converters. These devices are very expensive, and very large in size than similar AC devices like CBs. The HVDC rectifier and HV DC DC-DC converters use IGBTs and IGCTs, which are awfully expensive. There is a four terminal HVDC system in India and a five-terminal system in China, and more are planned. MTD has low power losses, high power throughput, smaller ROW, and offshore long cables may avoid congestion.

HVDC System Aspects – VSC vs LCC

HVDC converter systems are designed using two different topologies:

1) Self-Commutated Voltage Source Converter (VSC) semiconductor switches using IGBTs or IGCTs
2) Line Commutated Current source converter (LCC) based on thyristor-controlled switches with turn-on capability.

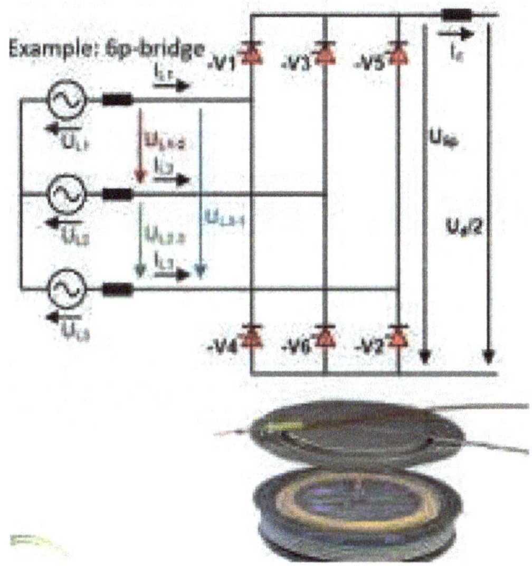

Figure 8.6.8 A typical LCC Converter Technology Topology

Modern HVDC LCC installations use 12-pulse converter bridges – i.e., Two six pulse bridges in series to cancel majority of harmonics. This thyristor technology consists of

1) Electrically triggered thyristors (ETT)
2) Direct Light-triggered Thyristor (LTT)

Typical ratings for the above are:

1) Thyristor wafers available in a typical diameter size are as follows:

4” diameter for DC currents up to 22,200 amps

5” diameter for DC currents up to 4,000 amps

6” diameter for DC currents up to 6,000 amps

While maximum Bipole ratings are 8,000 to 10,000 MW.

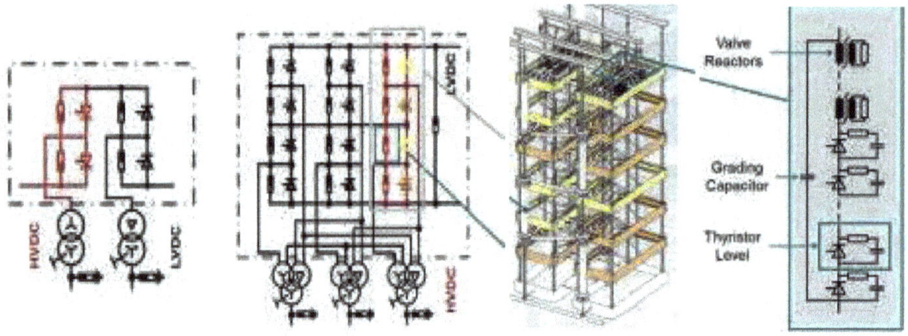

Figure 8.6.9 A typical LCC Converter Topology with Thyristor valves

In the LCC conversion technology the thyristor valves can have extremely high DC voltages and power Ratings Long year experience, and high degree of maturity High inherent overload capabilities Robustness with respect to DC side faults with fast clearing and recovery circuits Low power losses.

LCC Converter Topology Principle Characteristics

Figure 8.6.10 A typical LCC Converter Topology

I_d can only flow in one direction. Magnitude of P/I_d is controlled by the DC voltages Where

$$I_d = \left(U_{d\,rectifier} - U_{d\,inverter}\right\}/R \qquad (8.6.1)$$

Direction of P is controlled by changing the polarity of the DC voltages ($U_{d\,rectifier}, U_{d\,inverter}$)

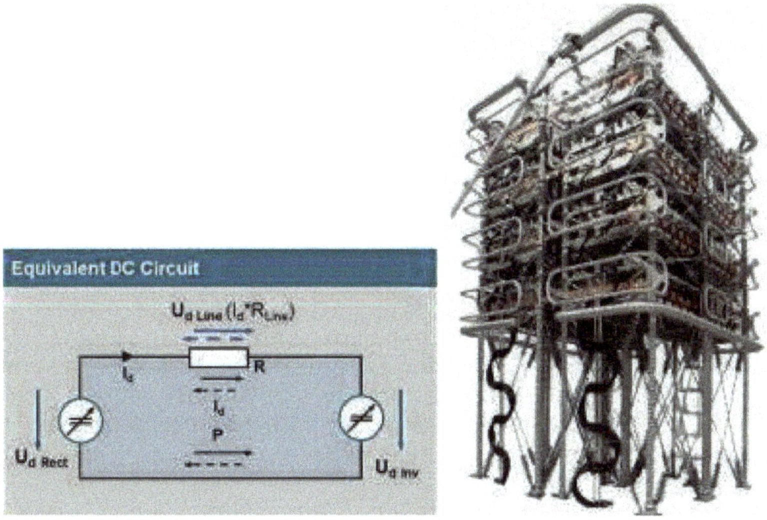

Figure 8.6.11 A typical VSC Converter Topology

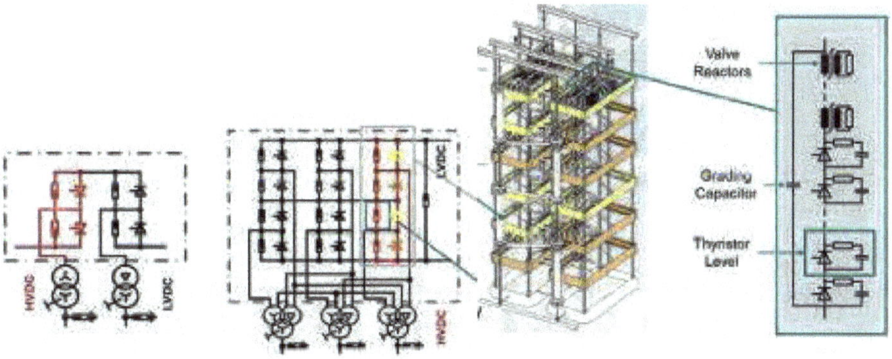

Figure 8.6.12 A typical LCC Converter Topology with Thyristor valves

LCC with converter technology with thyristor valves is used for extremely high DC voltages and power ratings. This system has long years of experience and high maturity, with high inherent overload capacities. Such LCC systems are robust with respect to DC side faults and have fast fault clearing and recovery. Such systems also have low power losses.

VSC Converter Technology

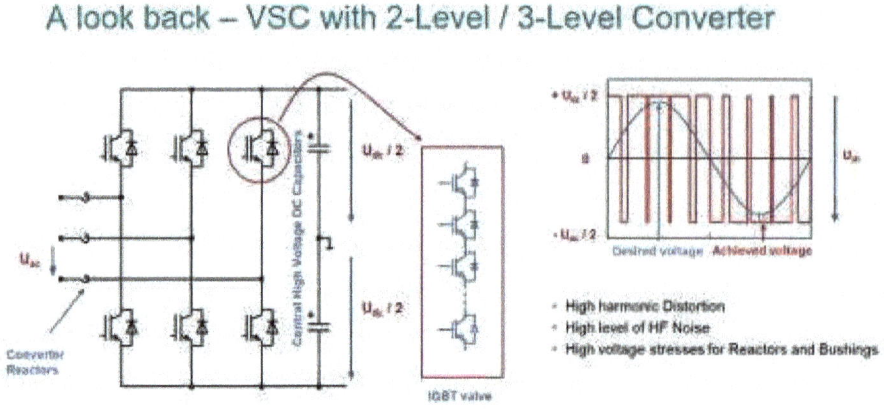

Figure 8.6.13 A typical VSC Converter Topology with 2 level 3 level converters

The above is VSC with a 2-level/3-level converter. This system gives rise to high harmonic distortion, high-level of high-frequency noise, and high voltage stresses for reactors and bushings.

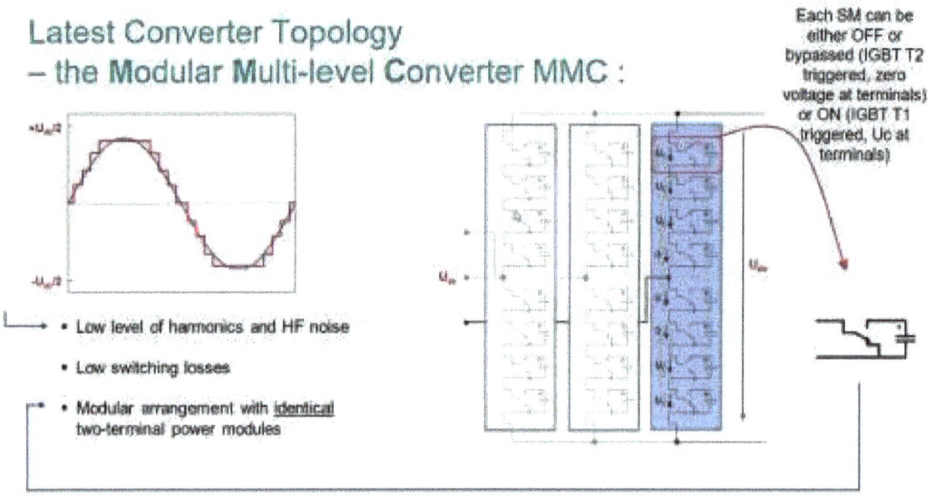

Figure 8.6.154 VSC Converter Topology and its performance

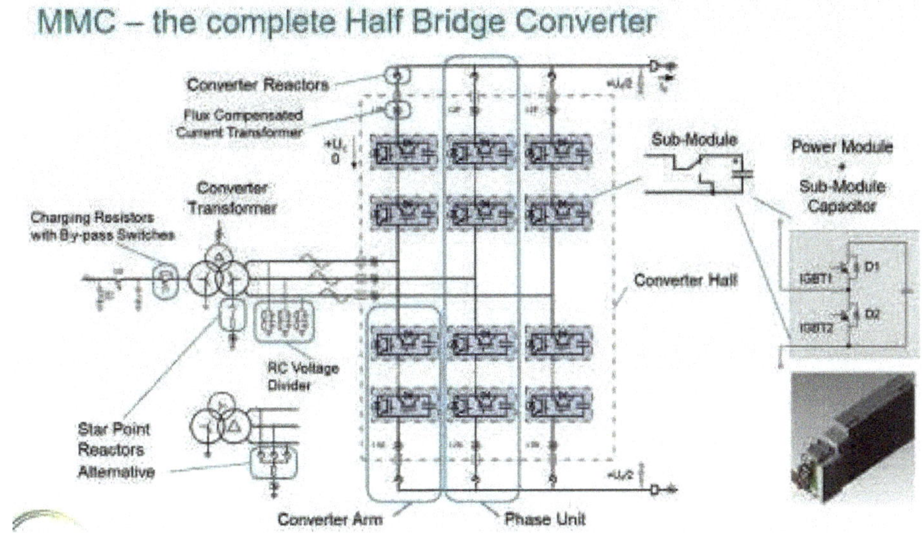

Figure 8.6.15 VSC Converter Topology with complete Half Bridge converter configuration

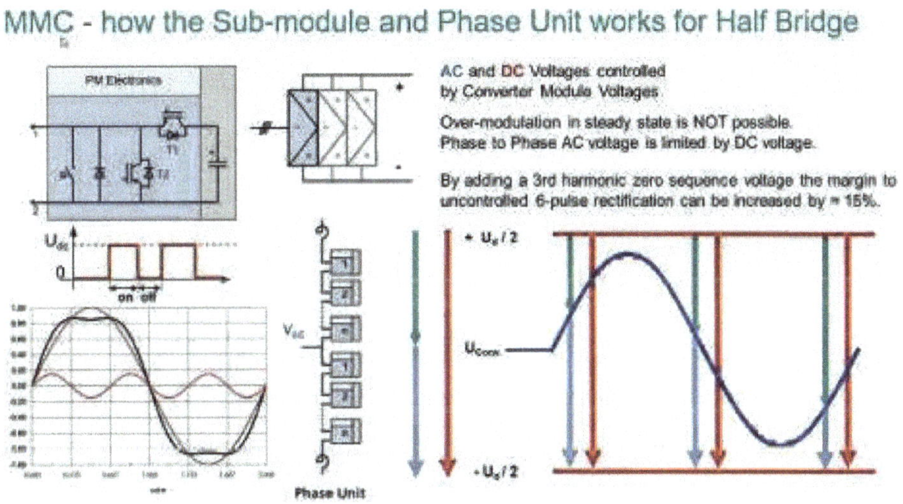

Figure 8.6.16 VSC Converter Topology with complete Half Bridge converter configuration performance of output voltage

Figure 8.6.17 Example of VSC Converter Topology with complete Half Bridge converter configuration

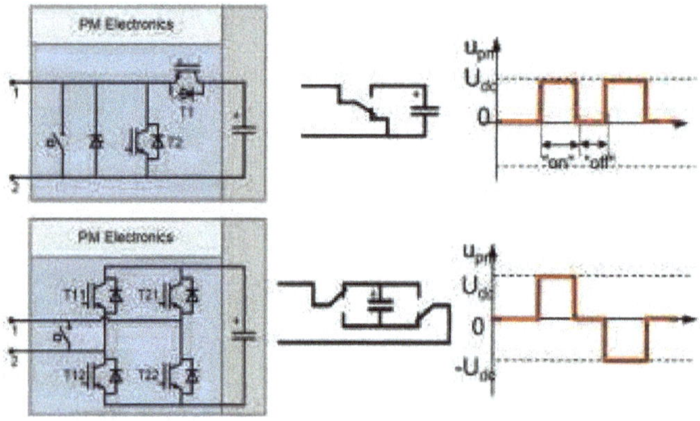

Figure 8.6.18 VSC Converter Topology with complete Full Bridge vs Half Bridge converter configuration performance comparison

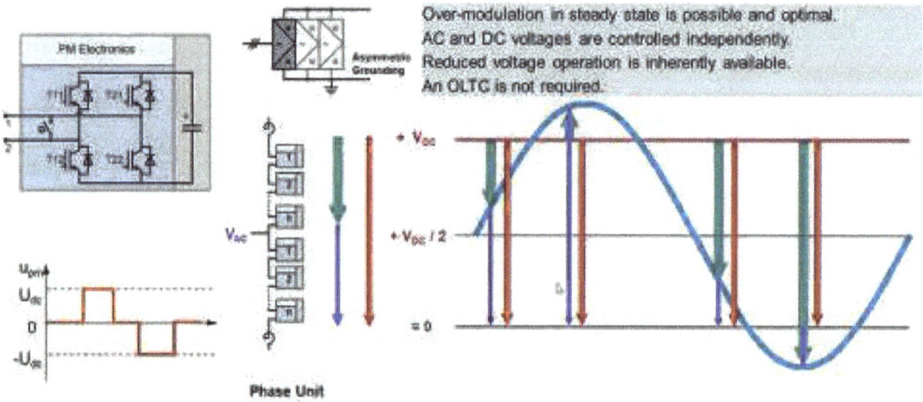

Figure 8.6.19 VSC Converter Topology with complete Full Bridge converter configuration and performance of sub-module and phase unit

Circuit Breakers and Switches like IGBTs and IGCTs used in such HVDC Systems

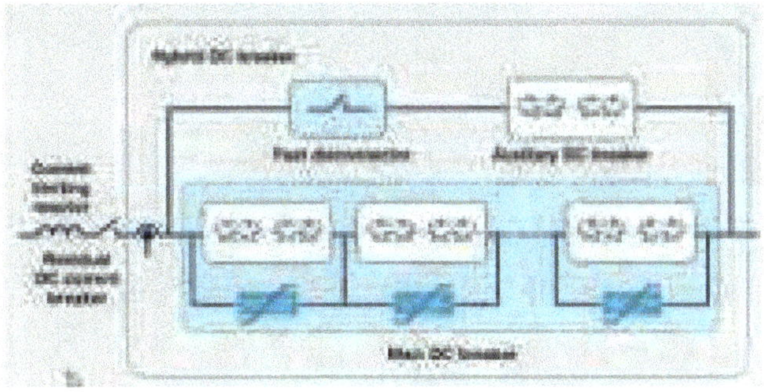

Figure 8.6.20 DC Breakers for HVDC System

HVDC System with DC Choppers as in Figure 8.6.21 below.

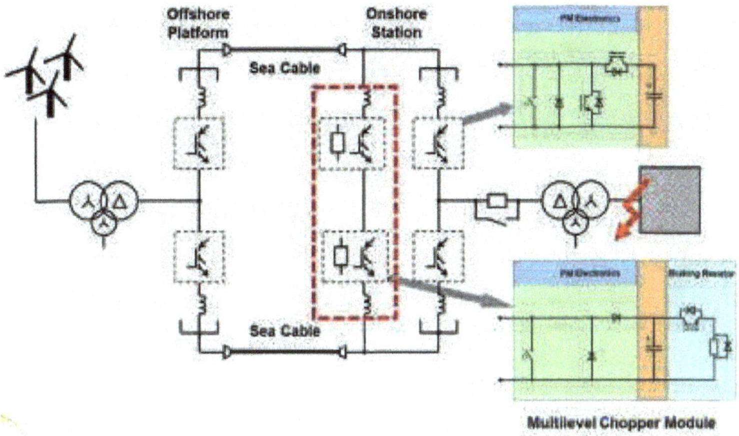

Figure 8.6.21 HVDC System with DC Choppers and their sb-modules

HVDC Systems and LCC Converters

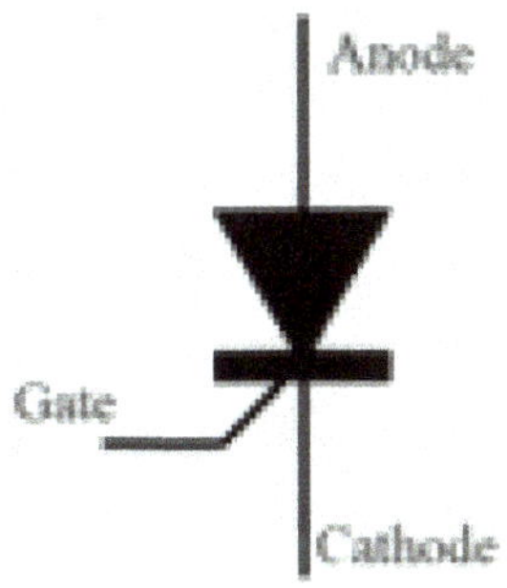

Figure 8.6.22 Simple LCC Converter in the form of thyristor

Figure 8.6.23 Typical high-power thyristors for use as HVDC valve

Figure 8.6.24 LLC HVDC valve configuration

Figure 8.6.25 Old Mercury-based LCC Converters

The system is shown below in Figure 27:

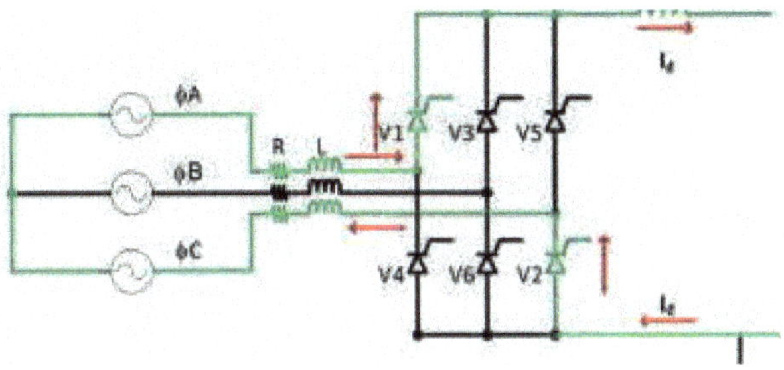

Figure 8.6.26 LCC Converter circuit configuration

Further circuits are shown for different configurations as below:

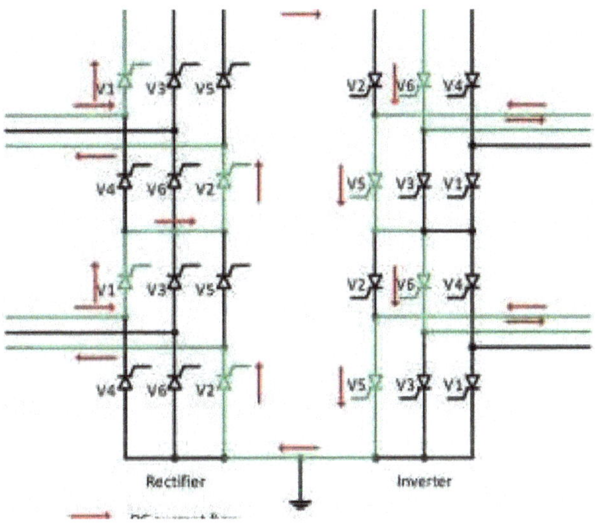

Figure 8.6.27 LCC Converter configuration and operation

Another LCC Converter configuration setup is shown below in Figure 29:

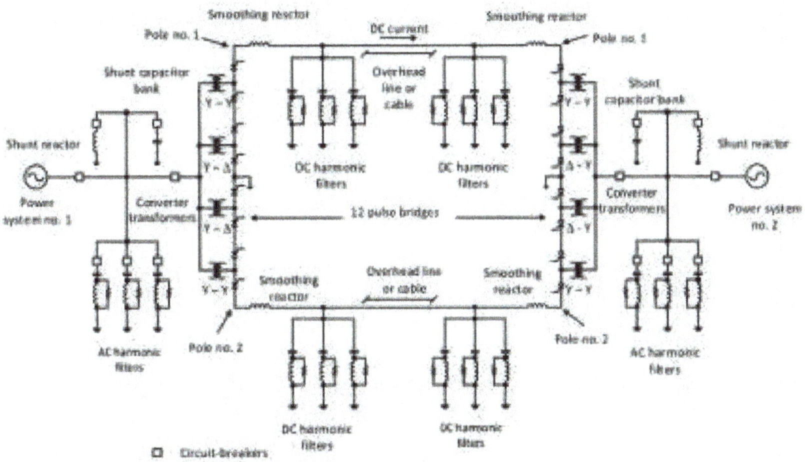

Figure 8.6.28 Alternate LCC Converter configuration

Each arm of the LCC converter has several series-connected thyristors. Thus, for a 500 kV HVDC converter there could be 70-90 thyristors connected in series as required per arm of the six pulse bridges. The number of series-connected thyristors depends on the voltage withstand capacity of each thyristor, which is generally between 5-6 kV. The insulation coordination requirements, along with basic impulse level withstand capability is also important.

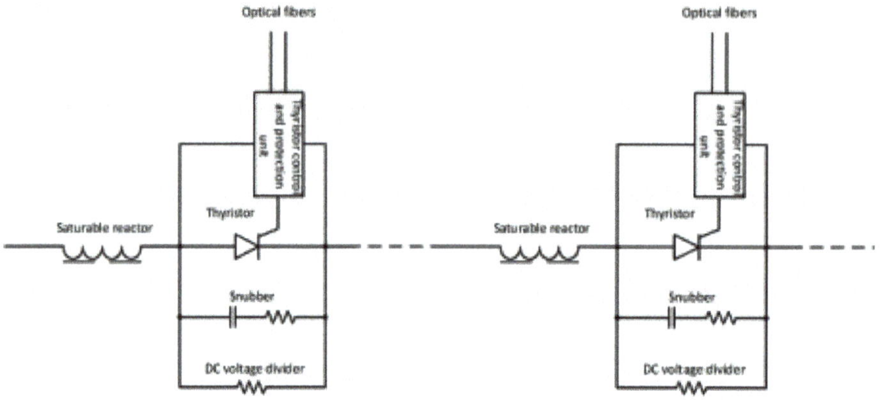

Typical layout of a LCC converter arm and thyristor levels

Figure 8.6.29 Further LCC Converter arm and thyristor levels configuration

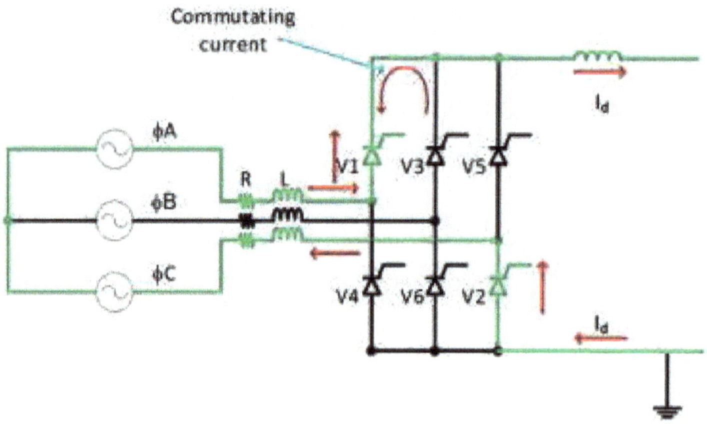

Figure 8.6.30 LCC Converter with operation of valves indicating the firing sequence below

In LCC converters, each valve arm will conduct current for a duration of $120^0 + \alpha$, where α is the overlap angle. When valve $V_1$is conducting the DC current, the next valve to be fired on the upper part of a six-pulse converter is V_3. When valve $V_3$is triggered, a phase-to-phase short circuit is established in the loop formed by V_1 and V_3. This short circuit current generates a current zero for V_1. Later, the current in $V_1$is interrupted, and the DC current is transferred to valve V_3.

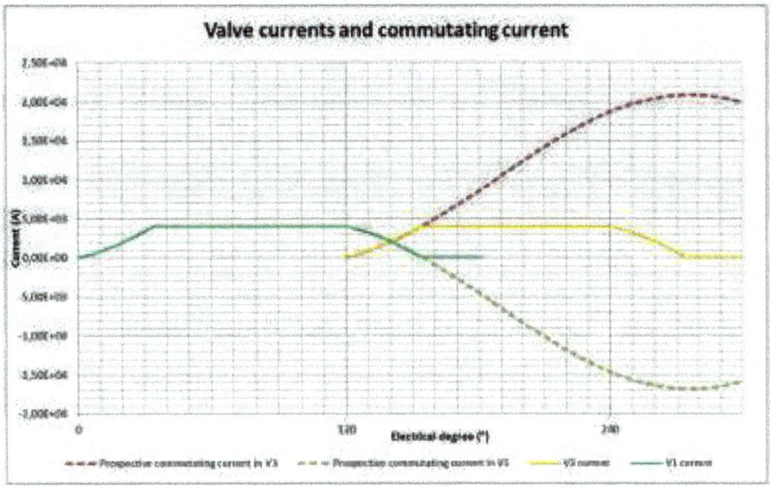

Figure 8.6.31 Performance of LCC Converters, valve currents, and commutation currents

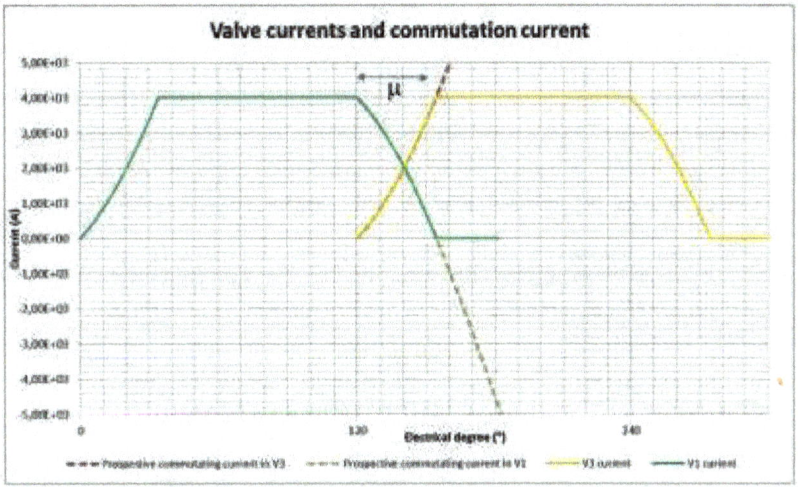

Figure 8.6.32 LCC valve currents and commutation currents

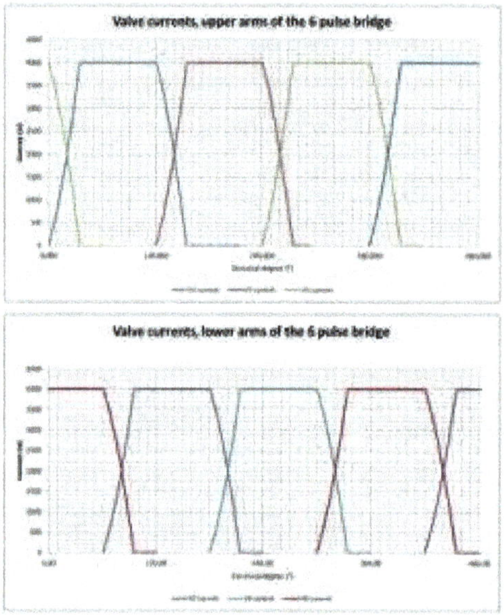

Figure 8.6.33 LCC valve currents and commutation currents continued

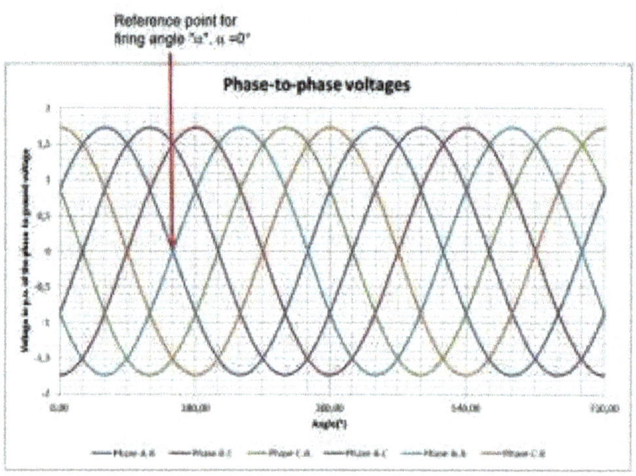

Figure 8.6.34 LCC converter output voltage

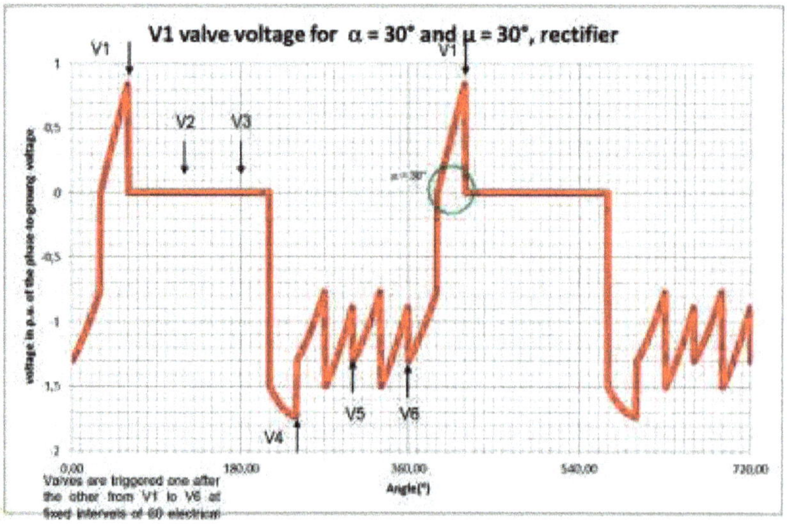

Figure 8.6.35 LCC valve voltage

VSC Converters

IGBTs are used is VSC converters with a small voltage applied to the gate, and removing it can control the ON and OFF conditions for successful operation at high voltage, resulting in self commutation unlike the LCC converters which need the help of the AC power.

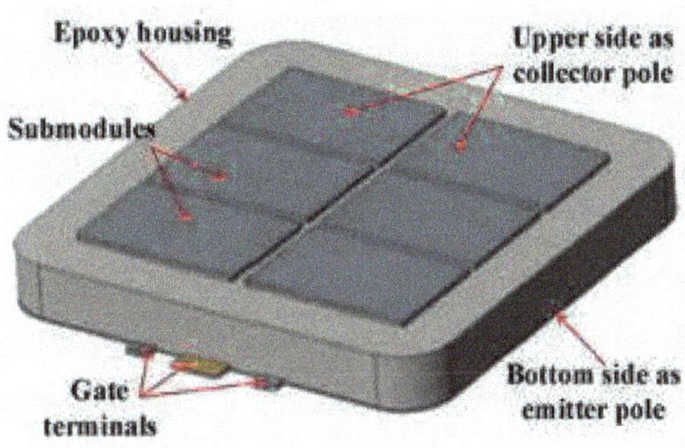

Figure 8.6.36 High Power IGBT module as a valve for use in HVDC system

A typical VSC HVDC valve system is shown in Figure 38 below:

Figure 8.6.37 A typical VSC HVDC system

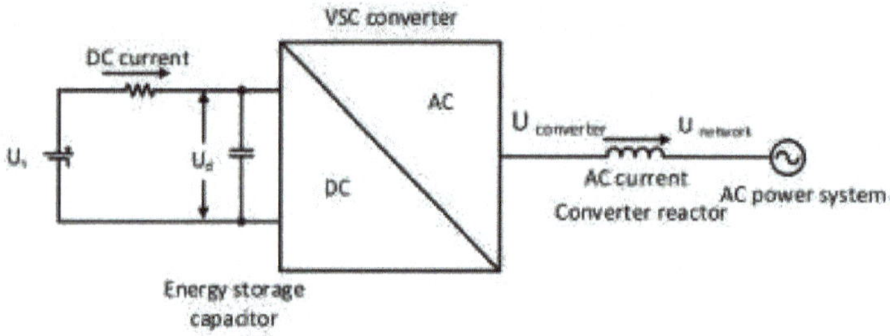

Figure 8.6.38 VSC converter system

Depending on the voltage amplitude and phase relationship of the voltage generated by the VSC valve in comparison with the AC network voltage, the VSC converter can act as:

1) A rectifier taking active power from AC network to the DC network.
2) AN inverter taking power from DC network to the AC network.
3) A STATCOM absorbing or generating reactive power to the AC network.

Active and reactive power transfer for VSC converter can be determined by the following equations:

$$P = \frac{U_{AC\ Network}\ x\ U_{converter} X\ sin\delta}{X}$$

$$Q = \frac{U_{AC\ Network}\ x\ (U_{AC\ Network}\ - U_{converter}\ x\ cos\delta)}{X}$$

Where,

$U_{AC\text{-}Network}$= AC system voltage

$U_{converter}$= Converter voltage

δ = Phase difference between AC network voltage and converter voltage

X = total reactance between VSC system and AC system; which is typically the sum of the reactance of converter reactor and AC system transformer

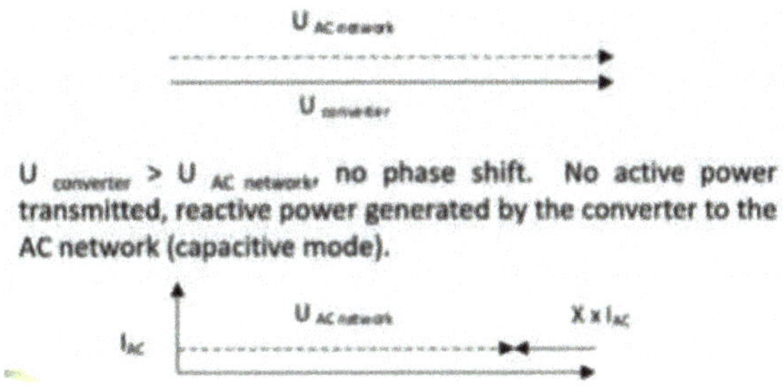

Figure 8.6.39 VSC converter system phasor diagrams for different power factors

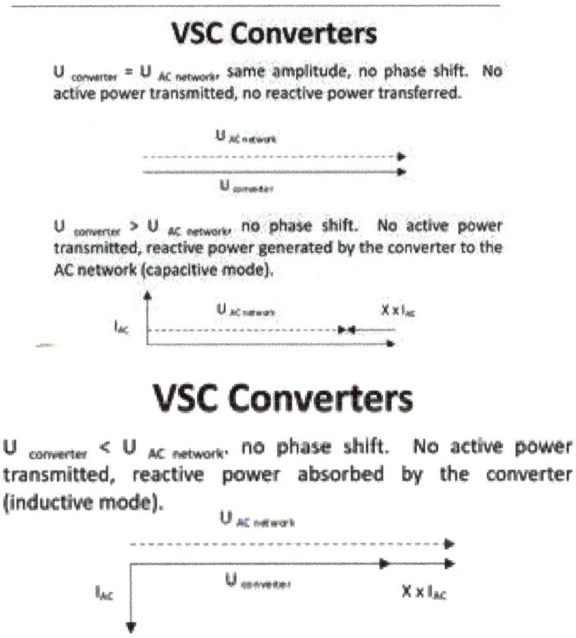

Figure 8.6.40 VSC converter system phasor diagram at upf

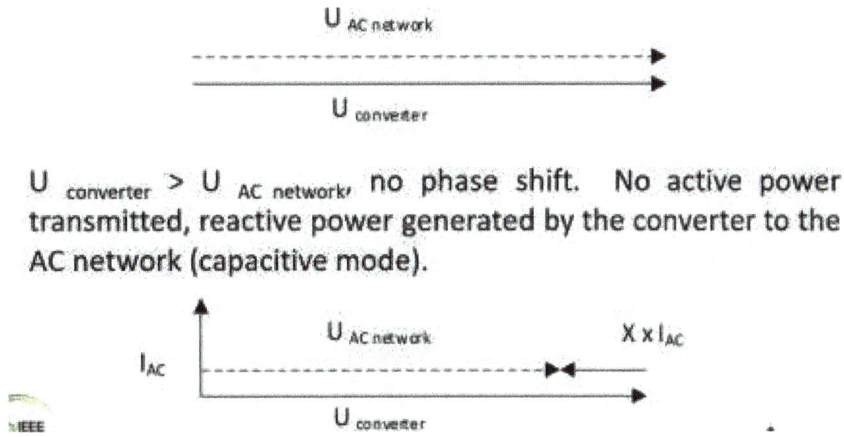

Figure 8.6.41 VSC converter system phasor diagram at leading pf

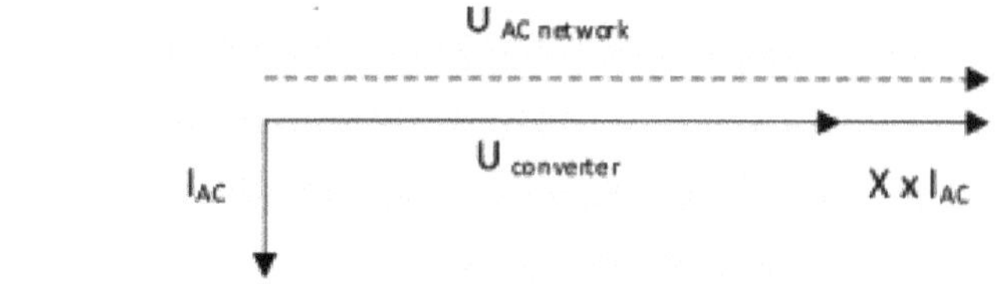

Figure 8.6.42 VSC converter system phasor diagram at lagging pf

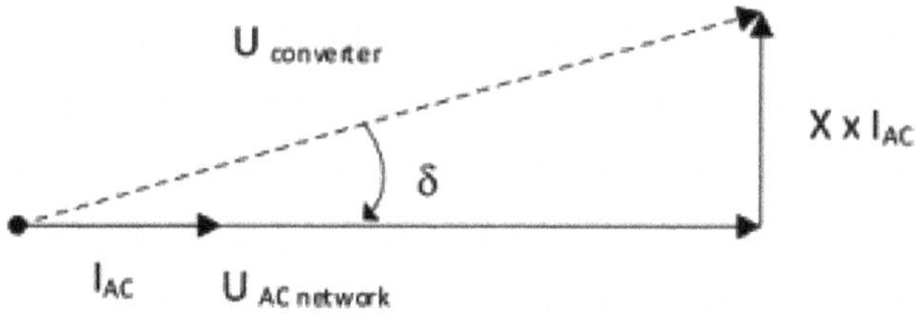

VSC Converters

They are two main groups of VSC converters:

- VSC converters of a "switch" type (past technology).
- VSC converters of a "controllable voltage source" called also MMC ("Modular Multi-level converter"), today's most offered technology.

VSC Converters

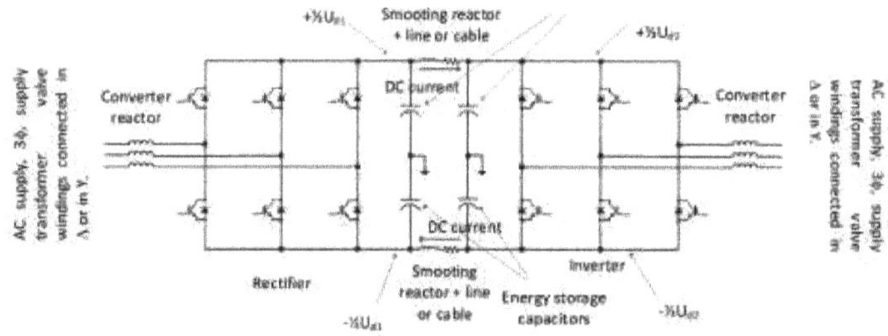

Typical arrangement of a two level VSC converter of a switch type

Figure 8.6.43 VSC converter of a switch type

A typical arrangement of a two-level VSC converter of a switch type is shown in Figure 43 above.

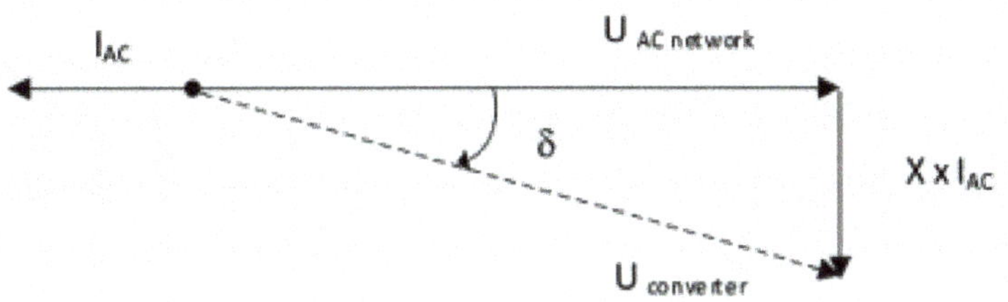

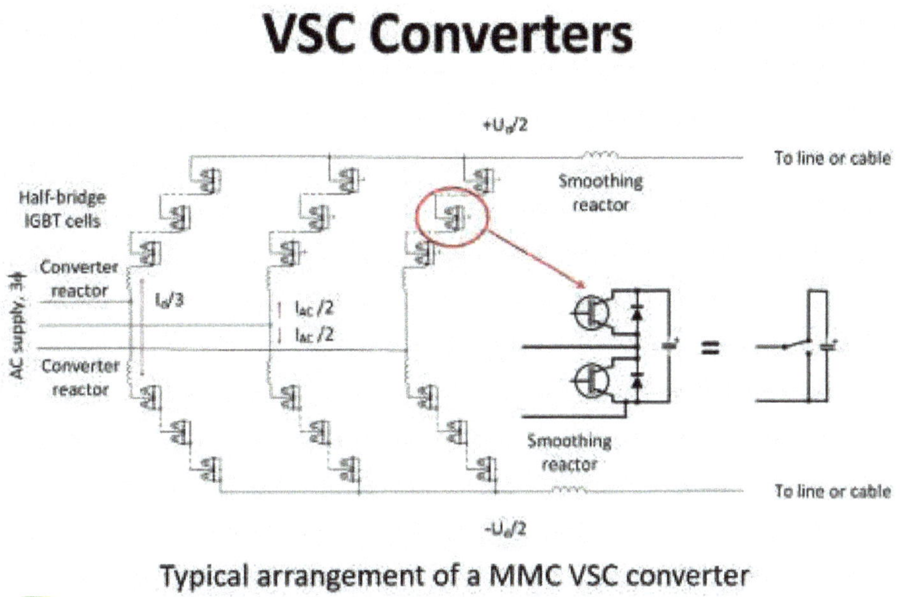

Figure 8.6.44 VSC converter system of MMC type

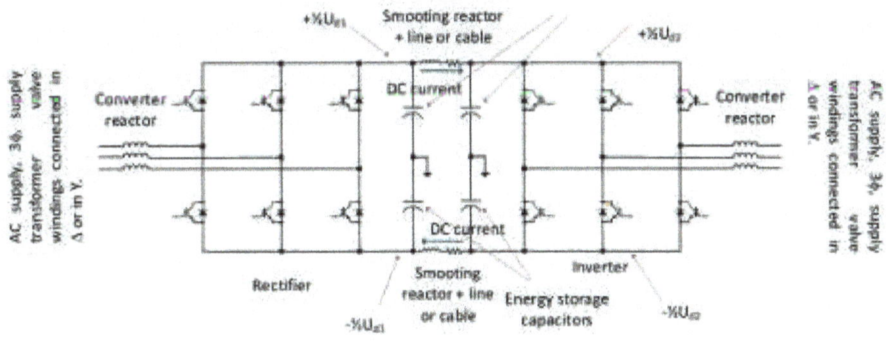

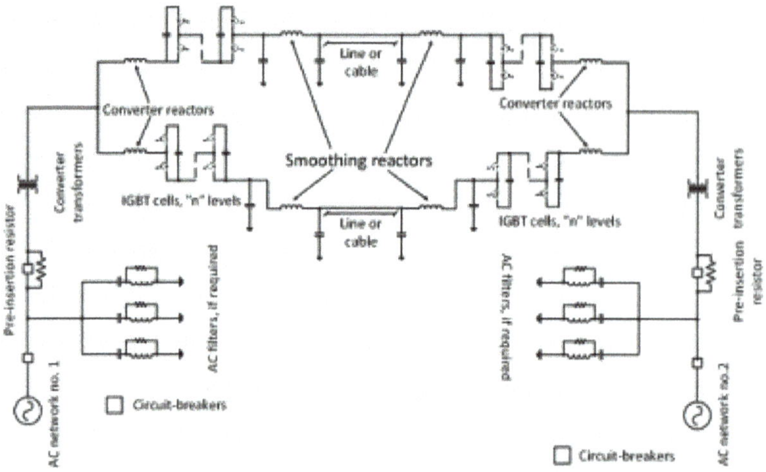

Figure 8.6.45 VSC converter system terminal connection layout at transmission end

A typical monopolar VSC converter transmission layout with surge arresters and instrument transformers are not represented.

There are different kinds of VSC converters. The switch-type converters do need external energy storage capacitor. These capacitors are at each cell level. Thus, for a typical +-345 kV DC voltage converter, approximately 230-325 cells/arm are necessary. The number of series-connected cells depends on the voltage capability of each cell and insulation coordination requirements.

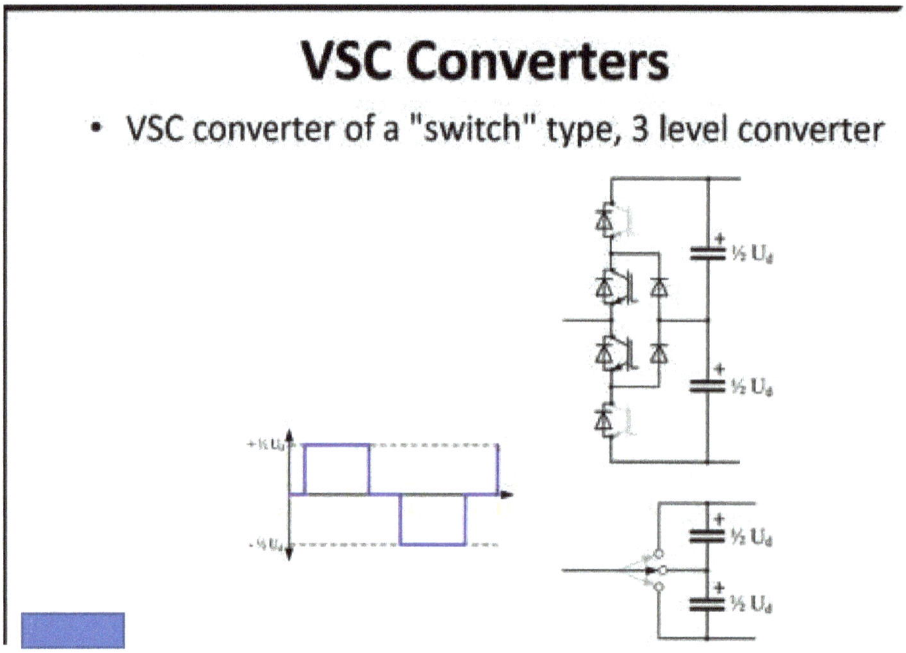

Figure 8.6.46 VSC converter system of a switch type for a 3-level converter and its performance

DC Circuit Breakers (CB)

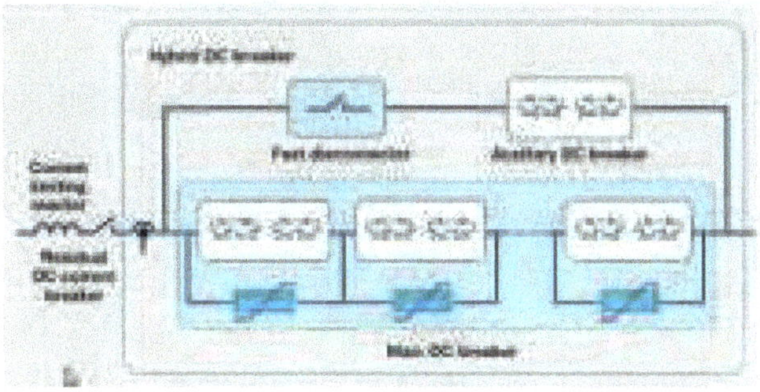

Figure 8.6.47 DC Breakers for HVDC system

To ensure fast and selective DC fault clearing in meshed DC grids, DC breakers have been developed and breaker demonstrators and suitable prototype CBs have been tested in field. Such DC breakers have to:

1) Provide an artificial current zero
2) Absorb the vast amount of energy stored in the DC system
3) Provide much smaller fault current interruption time (a few mili seconds) than conventional AC CBs
4) Require large space compared to AC CB and may need to be protected from an environmental condition.

HVDC System with DC Chopper

A typical HVDC system with a DC chopper is shown in Figure 48 below. The subsystem modules are shown in the inscts.

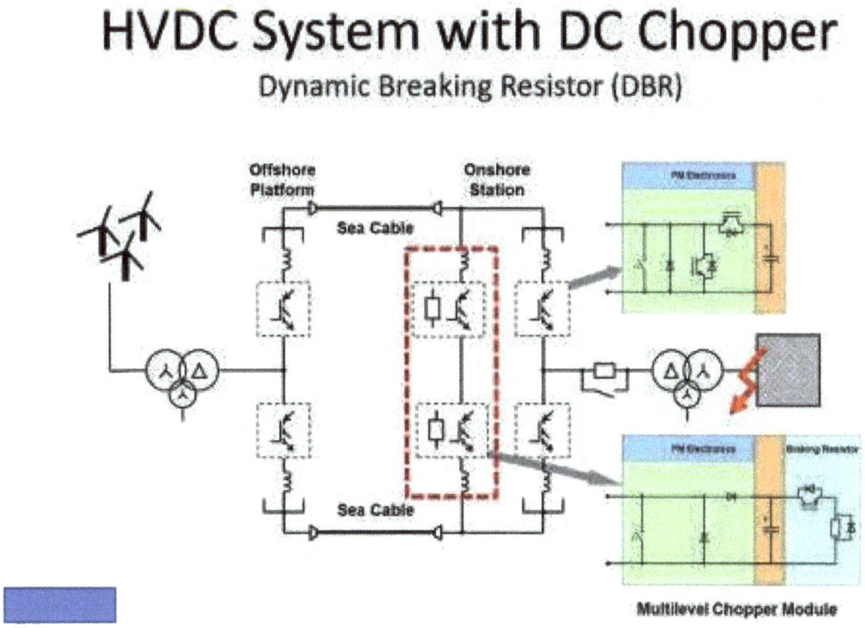

Figure 8.6.48 HVDC system with a DC chopper

VSC converters of switch type always need an external storage capacitor. Controllable voltage source converters have their own energy storage at each cell level. For +- 345 kV DC voltage converter, approximately 250-275 cells/arm are required. At the same time, the series-connected cells depend on the voltage capability of each cell and insulation coordination requirements.

HVDC System Design

Introduction

Tailor-made designs of HVDC systems and transmission links worldwide are tailor-made. There is therefore need for extensive system studies for all links, which have been demonstrated in Chapter 8, example presented in the prequel.

Control system is necessary for HVDC links that are actively linked, and the design and tuning of the control system is also an important part of the system design.

AC Grids

AC Grids are illustrated in Figure 49 below. Operational constraints in AC Grids are shown below in Figure 50 below:

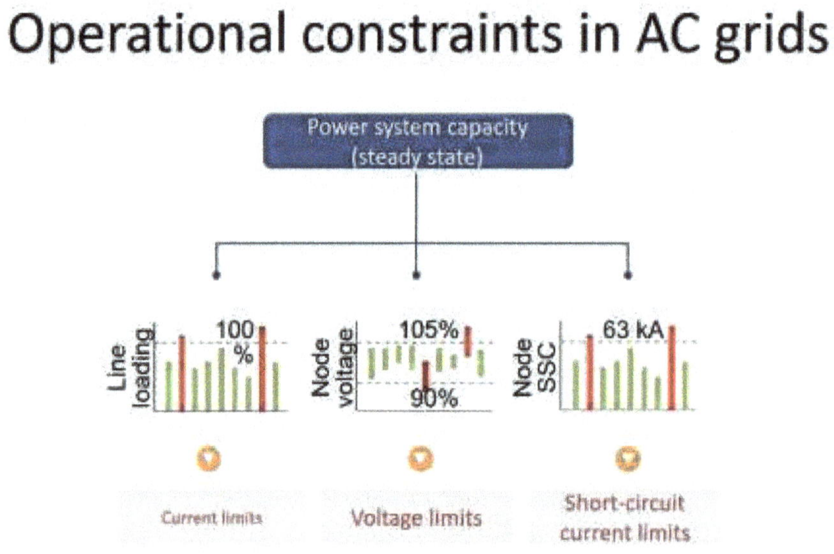

Figure 8.6.49 AC Grids

Figure 49 shows the constraints laid down by different load conditions. These pose tremendous operational constraints on the AC grid.

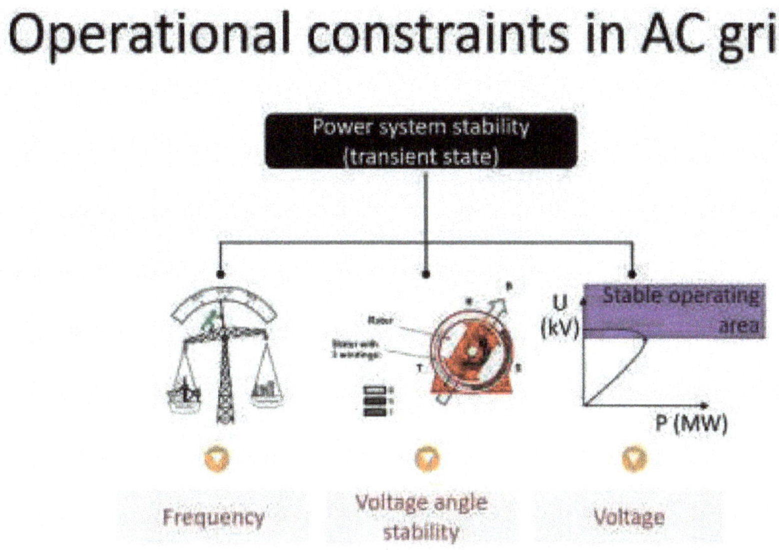

Figure 8.6.50 Operational constraints in AC Grids

The operational constraints in AC grids are associated with frequency, voltage angle stability, and voltage itself in relation to power system stability with respect to transient conditions and steady-state performance.

Dynamic Performance testing

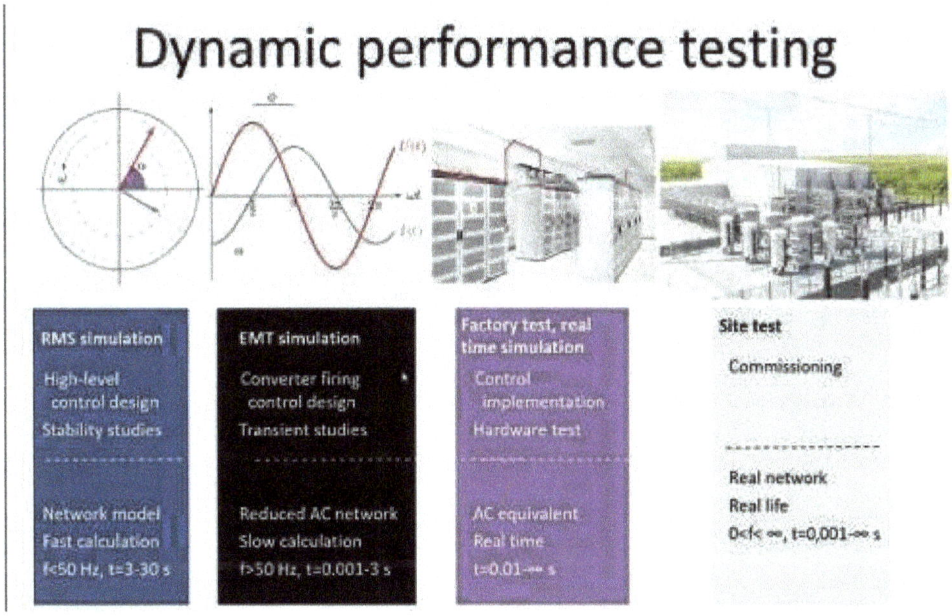

Figure 8.6.51 Dynamic performance testing

Firstly, this Dynamic performance testing involves RMS simulation for high-level control design and stability studies. Network modeling is a must for fast calculations for f < 60 Hz at t = 3-30 secs.

Secondly, one conducts an EMT simulation, which involves a converter firing sequence and control design.

Followed by transient studies. For this to happen, one needs reduced AC network, and slow calculations at f>60 Hz.

Thirdly, one conducts the factory tests followed by real-time simulation, verifying control design implementation and hardware tests. This, for an AC equivalent circuit, happens at t = 0.01 to infinity seconds.

Finally, site simulation and commissioning for f =1 to infinity and time = 0.01 to infinity.

In summary, all the above steps will enable the designer to verify the design process.

Design Tools

1) Calculations
a. Main Circuit Calculations
b. Converter Station loses
2) Electromagnetic Transient Simulations (EMT) in time domain
a. Transient over-voltages
b. Dynamic performance studies
c. Sub-Synchronous Torsional Interactions (SSTI)
3) Dynamic Vector Simulation (RMS in time domain)
a. Power Oscillation Damping
b. Frequency Control
4) Real-Time Simulations (in time domain)
a. Factory System Test of control equipment
b. Use in dynamic studies
5) Harmonics Simulation in frequency domain
6) CAD
a. Single Line Diagram (SLD)
b. Mechanical Design

Input to System Studies

1) Required data for the HV data links, i.e., active, and reactive power
2) Data of the connected active networks, i.e., Voltage levels and insulation coordination
3) Input for harmonic studies, i.e., limits and impedance sectors of harmonics
4) Data of the DC links, i.e., resistance, inductance, and capacitance
5) Network data for dynamic studies, i.e., normally model with MS model
6) Dynamic requirements, i.e., step response, settling time, overshoot (OS), and start-up times due to external faults
7) Data of nearby system links; facts of devices and generators
8) Physical layout limitations of the converter stations

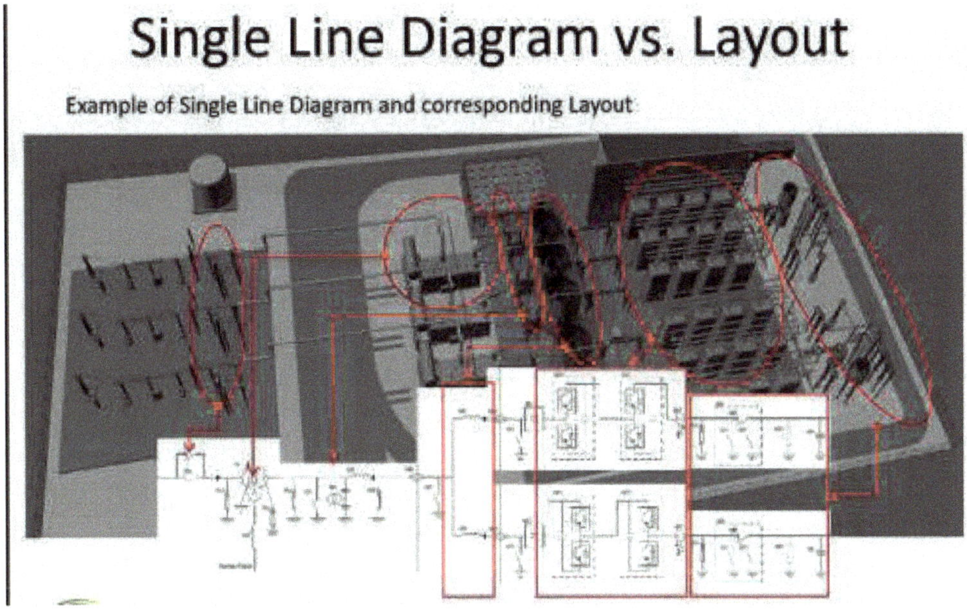

Figure 8.6.52 SLD and layout of systems (see reactive redlines showing circuit component configuration)

Requirement of Specifications for main circuit equipment of an HVDC system

1) Bill of quantity and spares
2) Reference to standards
3) Rating data, voltages, currents, impedance

a. e.g. Data comes from

i. System Studies
ii. Technical Specification
iii. Standards that need to be referred to

4) Accessories like Gas relays, WTIs, OV/UV relays etc.
5) Required tests and test levels
6) Information to be given by sub-suppliers, like losses

Chapter 9
Summary

The study of DC-DC Converters for HVDC Transformations is very intriguing and challenging. In this book I have tried to do some justification to several topics of interest, starting with the description and necessity of DC-DC Converters. This is followed by describing different block diagrams of several DC-DC Converters like:

1) Buck Converters – when output voltage is less than input voltage in magnitude
2) Boost Converters – when the output voltage is greater in magnitude than the input voltage
3) Buck-Boost Converters – when either output voltage is less than input voltage in magnitude or when output voltage is greater in magnitude than the input voltage
4) SEPIC Converters – when the output voltage is greater in magnitude than the input voltage or when the output voltage is less than the input voltage in magnitude or when output voltage is equal to the input voltage in magnitude. Besides, the output of this converter is decided by the duty cycle of the control switch.
5) Fly back Converters – These are used both in AC/DC or DC/DC conversion and have a galvanic isolation between the input and output magnitudes of voltages.
6) Direct Converters – describing the finer nuances of direct converters succinctly described in above items.
7) Indirect Converters - describing the finer nuances of direct converters succinctly described in above items.
8) After understanding all these different converters, we described the filters used for such converters followed by a Case Study that describes a project with LVDC/MVDC/HVDC for an On-Shore Wind Power substation. Similar case studies can be researched and developed for Solar plants, and other alternate energy plants in the LVDC/MVDC/HVDC using DC-DC Converters for HVDC Transformations as elaborated in this text. Station, as in Chapter 8. An additional case study has been added after my recent travels to Grimsel - II hydroelectric plant in the town of Innertkirchen, Switzerland in February of 2020, just before the harrowing COVID-19 pandemic struck the entire world.
9) Finally, a comprehensive summary is provided in Chapter 9 to assist instructors in dividing the coverage of the above topics into two-semester courses as per the requirement of instructions to senior undergraduate students or incoming graduate-level students, respectively.

References

[1] R. D. Middlebrook, "Design Techniques for Preventing Input Filter Oscillations in Switched-

Mode Regulators," Proceedings of the Fifth National Solid-State Power Conversion Conference, Powercon 5, pp. A3-1 through A3-16, May 1978.

[2] ISSN: 2278-0181 Design of Buck-Boost Converter Using Multisim Software. Mousumi Mishra, Assistant. Professor, KIIT University.

[3] "Soft-switching characterization of medium voltage IGBT power modules and press-pack diodes in a 1kHz mega-watt dc/dc resonant converter", C. Dincan, P.C.K jaer, S.M. Nielsen, C.L. Bak, Proc. IEEE Eur. Conference Power Electron and Application. (EPE2018).

[4] " Design of a high power, resonant converter for DC wind turbines", C. Dincan, P.C.Kjaer, Y. Chen, S.M.Nielsen, C.L.Bak, E. Sarra, V. Sriram – Trans. In Power Electronics and Applic., *under review*

[5] "Control Strategies for Buck/Boost Converters in Electric Vehicle Applications", Esmaeil Gandomkar and Hemchandra Shertukde; to be presented at an appropriate conference in 2026

Index

D

DC-DC Converters: Power converters that change one DC voltage level to another. [*Pg.: 1, 3, 5, 9, 11, 14, 15, 24, 45, 66, 69, 70, 76, 116, 117, 125, 137, 144, 145, 149, 150, 173*].

DCM (Discontinuous Conduction Mode): A mode of operation where the inductor current falls to zero during part of the switching cycle. [*Pg.: 43, 62, 63, 95*].

Diode Losses (VD, Qrr): Voltage drop (VD) and reverse recovery charge (Qrr) causing energy loss in diodes. [*Pg.: 111*].

Duty Cycle (D): The fraction of one switching period in which a signal or system is active. [*Pg.: 24, 31, 34, 36, 37, 38, 40, 41, 42, 47, 48, 49, 61, 117, 173*].

E

Efficiency (η): The ratio of output power to input power, representing performance. [*Pg.: 4, 11, 29, 41, 45, 46, 47, 48, 53, 57, 59, 61, 62, 66, 69, 70, 79, 81, 83, 86, 88, 90, 94, 143, 144*].

EMI (Electromagnetic Interference): Noise generated by electronic devices that may disrupt operation. [*Pg.: 9, 10, 69, 81, 83, 127*].

ESR (Equivalent Series Resistance): Internal resistance of capacitors causing power loss. [*Pg.: 56, 60, 70*].

F

Flyback Converter: A DC-DC converter that stores energy in an inductor during the switch-on phase and transfers it to the output during the switch-off phase. [*Pg.: 56, 143*].

Filters (LC, EMI): Circuits that remove unwanted frequencies or smooth voltage/current waveforms. [*Pg.: 2, 15, 28, 57, 69, 70, 71, 173*].

Frequency Control: Regulation of the switching frequency in converters for stability and performance. [*Pg.: 59, 91, 171*].

G

Galvanic Separation: Electrical isolation between different parts of a circuit for safety. [*Pg.: 77, 87*].

Gate Drive: A circuit that provides appropriate voltage/current to turn on/off a transistor such as a MOSFET or IGBT. [*Pg.: 46, 47, 48*].

H

Half Bridge / Full Bridge Converters: Power converter topologies using two or four switching devices respectively. [*Pg.: 143, 154, 155*].

Harmonic Simulation: Analysis of harmonic components in electrical signals. [*Pg.: 137, 150, 153, 171*].

HVDC (High Voltage Direct Current): A system for transmitting DC electricity at high voltages over long distances. [*Pg.: 1, 2, 14, 69, 76, 81, 88, 116, 117, 118, 137, 144, 145, 146, 148, 149, 150, 155, 156, 157, 159, 162, 163, 169, 172, 173*].

I

IGBT (Insulated Gate Bipolar Transistor): A high-efficiency switch used in power electronics. [*Pg.: 1, 2, 83, 90, 94, 107, 108, 109, 110, 111, 145, 148, 150, 155, 162, 174*].

IGCT (Integrated Gate Commutated Thyristor): A device combining thyristor switching with gate-controlled turn-off capabilities. [*Pg.: 1, 2, 83, 90, 145, 148, 150, 155*].

Input Impedance: The resistance seen by the source at the input terminals of a circuit. [*Pg.: 55, 70*].

IL1 / IL2: Inductor currents used in analysis and design. [*Pg.: 55, 56, 58, 59, 60, 61, 62*].

K

Kp, Ki, Kd: Constants for the proportional, integral, and derivative terms in a PID controller. [*Pg.: 19, 22, 27, 33, 89*].

L

L1 / L2 Inductors: Inductors used for energy storage and filtering in converters. [*Pg.: 7*].

LCC (Line Commutated Converter): A type of converter used in HVDC systems with thyristors. [*Pg.: 145, 146, 147, 150, 151, 152, 156, 157, 158, 159, 160, 161, 162*].

LLC (Resonant Converter): A soft-switching converter that reduces switching losses. [*Pg.: 87, 90, 142, 143, 157*].

Losses (Psw, Prl, PD1): Power losses due to switching, parasitic resistance, and diode operation. [*Pg.: 59, 60*].

M

MATLAB: A software tool for numerical computing and simulation, used for modeling converters. [*Pg.: 20, 22, 23, 27, 75, 105*].

MLC (Multilayer Ceramic Capacitor): A type of capacitor with high efficiency and small size. [*Pg.: 59, 62*].

MOSFET (Metal Oxide Semiconductor Field Effect Transistor): A switching device used in converters. [*Pg.: 2, 11, 12, 45, 46, 47, 48, 55, 59, 68, 125, 142*].

N

Negative Resistance: A condition where increasing voltage leads to decreased current, often undesirable. [*Pg.: 70*].

Noise Attenuation: Reduction of electrical noise in systems using filters or shielding. [*Pg.: 70*].**Overshoot (OS)**: When output temporarily exceeds the desired value during transitions. [*Pg.: 19, 171*].

P

Parasitic Components: Unintended resistances, capacitances, or inductances in a circuit. [*Pg.: 62*].

Peak Current: The maximum instantaneous current in a component. [*Pg.: 62, 63, 70, 133*].

PID Controller: A feedback loop mechanism for controlling processes using proportional, integral, and derivative components. [*Pg.: 18, 19, 23, 25, 27, 33*].

PSpice: A simulation tool for electronic circuit design. [*Pg.: 12, 25, 54, 125*].

Q

Qrr (Reverse Recovery Charge): Charge that must be removed from a diode during switching. [*Pg.: 96, 113*].

R

Rectifier: A device that converts AC to DC. [*Pg.: 1, 2, 62, 79, 82, 86, 87, 89, 90, 95, 102, 105, 107, 112, 113, 142, 150, 163*].

Resonant Converter: Converter topology that uses resonant circuits for efficient power conversion. [*Pg.: 79, 87, 91, 92, 98, 104, 108, 114, 174*].

Ripple Voltage (ΔVOUT): Undesired AC component in the output voltage. [*Pg.: 13, 125*].

S

SEPIC (Single-Ended Primary Inductor Converter): A converter topology allowing step-up and step-down operation. [*Pg.: 1, 11, 34, 42, 55, 56, 57, 58, 61, 66, 68, 75, 142, 173*].

SISO (Single Input Single Output): A control system with one input and one output.[*Pg: 17, 24*].

SMPS (Switched Mode Power Supply): A power supply using switching regulators for high efficiency. [*Pg.: 15, 28, 34, 56*].

Soft Switching: A method to reduce losses by switching devices under zero-voltage or zero-current conditions. [*Pg.: 94, 108, 112, 113*].

T

Thyristor: A four-layer semiconductor switch used in HV applications. [*Pg.: 2, 145, 150, 151, 152, 156, 157, 159*].

Transformer: A device that transfers electrical energy between circuits through magnetic induction.[*Pg.: 3, 6, 7, 31, 34, 56, 57, 68, 69, 78, 79, 80, 82, 83, 85, 87, 88, 89, 90, 92, 100, 101, 102, 103, 105, 111, 115, 118, 123, 124, 137, 143, 144, 148, 164, 167*].

Transition Losses: Energy lost during switching transitions in semiconductors. [*Pg.: 59, 61, 62*].

U

Under Voltage Lockout (UVLO): Protection circuit that disables the system when voltage is too low. [*Pg.: 46*].

V

VCC, VCE, VDS, VIN, VOUT: Voltage parameters used for analyzing circuits. [*Pg.: 4, 12, 13, 15, 26, 49, 55, 56, 57, 58, 59, 60, 61, 62, 97, 105, 117*].

VSC (Voltage Source Converter): A converter using semiconductor devices to convert DC to AC or vice versa. [*Pg.: 145, 146, 147, 150, 152, 153, 154, 155, 162, 163, 164, 165, 166, 167, 169*].

www.ingramcontent.com/pod-product-compliance
Lightning Source LLC
LaVergne TN
LVHW070853210926
848482LV00016B/928
9798349397028